# The Top 50 Mediterranean Island Plants

## Wild plants at the brink of extinction, and what is needed to save them

IUCN/SSC Mediterranean Islands Plant Specialist Group

**Edited by Bertrand de Montmollin and Wendy Strahm**

With assistance from Svenja Busse, Patricia Désilets and Marie Lafontaine

The designation of geographical entities in this book, and the presentation of the material, do not imply the expression of any opinion whatsoever on the part of IUCN concerning the legal status of any country, territory, or area, or of its authorities, or concerning the delimitation of its frontiers or boundaries.

The views expressed in this publication do not necessarily reflect those of IUCN.

| | |
|---|---|
| Published by: | IUCN, Gland, Switzerland and Cambridge, UK |
| Copyright: | © 2005 International Union for Conservation of Nature and Natural Resources |
| | Reproduction of this publication for educational or other non-commercial purposes is authorized without prior written permission from the copyright holder provided the source is fully acknowledged. |
| | Reproduction of this publication for resale or other commercial purposes is prohibited without prior written permission of the copyright holder. |
| Citation: | Montmollin, B. de and Strahm, W. (Eds). 2005. *The Top 50 Mediterranean Island Plants: Wild plants at the brink of extinction, and what is needed to save them.* IUCN/SSC Mediterranean Islands Plant Specialist Group. IUCN, Gland, Switzerland and Cambridge, UK. x + 110 pp. |
| Cover photos: | (From back cover left to right): *Euphorbia margalidiana* (Josep Lluis Gradaille); *Cheirolophus crassifolius* (University of Catania); *Astralagus macrocarpus* subsp. *lefkarensis* (Georgios Hadjikyriakou); *Bupleurum dianthifolium* (Lorenzo Gianguzzi); *Minuartia dirphya* (Gregoris Iatroú); *Anchusa crispa* (Federico Selvi); *Biscutella rotgesii* (Jean-François Marzocchi); *Salvia veneris* (Yiannis Christofides); *Centaurea gymnocarpa* (Bruno Foggi); *Limonium strictissimum* (Jean-François Marzocchi); *Centaurea akamantis* (Georgios Hadjikyriakou); *Delphinium caseyi* (Christodoulos Makris); *Calendula maritima* (Anna Giordano) |
| ISBN: | 2-8317-0832-X |
| Design, layout and produced by: | *Maximedia, Ltd.* |
| Printed by: | Information Press, Oxford, UK |
| Available in English and French from: | IUCN Publications Services Unit 219c Huntingdon Road, Cambridge CB3 0DL, United Kingdom Tel: +44 1223 277894, Fax: +44 1223 277175 E-mail: books@iucn.org www.iucn.org/bookstore A catalogue of IUCN publications is also available. |

The text of this book is printed on 100 gsm Fineblade Smooth which is made from 100% sustainable sources using chlorine-free processes.

# Contents

# Foreword

The Mediterranean region is renowned as one of our planet's main cradles of culture and civilization. However it is easy to forget that the region also houses an extraordinary natural heritage, which has resulted in it being identified among the 200 most important ecoregions in the world (Olson & Dinnerstein, 1998), as well as considered as one of the 34 global "hotspots" for conservation priority (Mittermeier *et al.*, 2004).

Throughout history the Mediterranean has inspired researchers who have catalogued its fauna and flora. While most birds and mammals found in the region also live elsewhere in neighbouring parts of Europe, Africa and Asia, this is not the case for the plant life of which many species are unique to the Mediterranean. The great number of islands in the region covering a wide range of altitudes has allowed new species to evolve as well as provide refugia for others to survive.

For thousands of years nature in the Mediterranean has been shaped by people, who have known how to use natural resources as well as develop their diversity. However for the past few decades this has not been the case. The introduction of intensive agriculture, infra-structure, urbanization and the development of mass tourism have profoundly changed living conditions. Skyrocketing population growth and climate change have exacerbated these changes, and the introduction of exotic plants has often eliminated the native species.

Today many Mediterranean plants are threatened with extinction. The disappearance of one species often brings about ecosystem changes which magnify the loss and further damage the quality of the environment. The goal of this booklet is not to make a complete list of these species, but rather to raise awareness about the extent of the problem and to illustrate progress.

It is not easy to mobilize the public and decision makers to take actions needed to save threatened species. It is hoped that the publication of this first "Top 50" booklet will con-tribute, alongside other tools such as the *IUCN Red List of Threatened Species*, to this goal.

I cannot conclude this foreword without thanking all the contributors, whose dedicated and meticulous work has allowed these case studies to be produced.

Luc Hoffmann,
President of the MAVA Foundation

# Mediterranean 'Top 50' islands and archipelagos

Columbretes     Corsica     Tuscan Archipelago

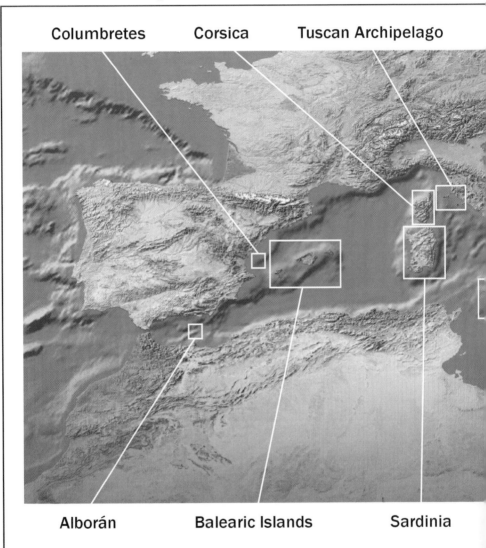

Alborán     Balearic Islands     Sardinia

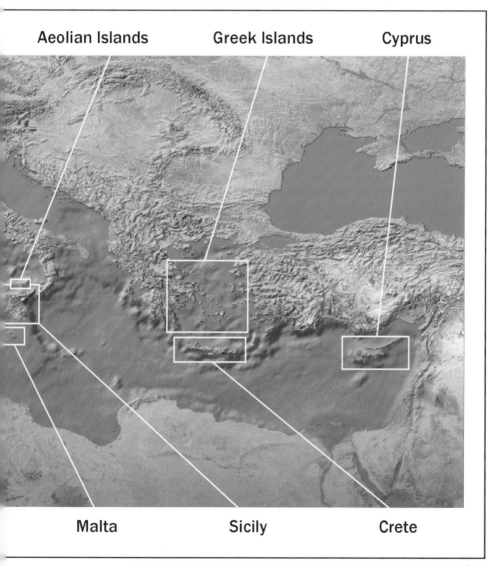

Aeolian Islands     Greek Islands     Cyprus

Malta     Sicily     Crete

*Background map courtesy of* **UNEP WCMC**

# Acknowledgements

The editors gratefully acknowledge the many collaborators who have made this project a reality, and apologise in advance to anyone who may have inadvertently been left out from this list.

This project was financed by the MAVA Foundation; in particular the strong support and encouragement from Luc Hoffmann and Pierre Goeldlin has been greatly appreciated.

Three IUCN interns have worked on this project: Patricia Désilets (July-December 2003), and Marie Lafontaine (October 2004-April 2005) who were supported by the Government of Québec International Governmental Organization Internship Program, and Svenja Busse, supported by the German agency BFIO – Bureau for International Organizations Personnel (February-July, 2004). Our warm thanks to them for their cheerful liaison with the many collaborators in this project.

The following scientists and conservationists have contributed their knowledge and photos to this project and are thanked for their patience (SSC members are marked with an asterisk): Gabriel Alziar* (Jardin Botanique de la Ville de Nice); Pier Virgilio Arrigoni (Università degli Studi di Firenze); Gianluigi Bacchetta (Università di Cagliari); Alfred E. Baldacchino (Malta Environment & Planning Authority); Gabriel Bibiloni (Universitat de les Illes Balears); Ignazio Camarda (University of Sassari); Charalambos S. Christodoulou (Forestry Department, Nicosia); Yiannis Christofides; Paolo Colombo; Theophanis Constantinidis (Agricultural University of Athens); Manuel B. Crespo (Universidad de Alicante); Olivia Delanoë* (INEA Ingénieurs-conseil, Nature, Environnement, Aménagements); Aljos Farjon*(RBG Kew); Eladio Fernández-Galiano (Council of Europe); Bruno Foggi* (Università di Firenze); Christina Fournaraki (Mediterranean Agronomic Institute of Chania, Kriti); Pere Fraga Arguimbau (Consell Insular de Menorca); Alain Fridlender (Université de Provence, Marseille); David Galicia Herbada (TRAGSA, Área de Medio Ambiente); Jacques Gamisans* (Université Toulouse III); Giuseppe Garfì (CNR – Istituto per i Sistemi Agricoli e Forestali del Mediterraneo (ISAFoM)); Lorenzo Gianguzzi (Università degli Studi di Palermo); Anna Giordano (WWF-Italia); César Gómez-Campo (Universidad Politécnica de Madrid); Josep Lluis Gradaille (Jardí Botànic De Sóller, Mallorca); Georgios Hadjikyriakou; Gregoris Iatroú (University of Patras); José Maria Iriondo* (Universidad Politecnica, Madrid); Armin Jagel; Ralf Jahn (Universität Regensburg); Daniel Jeanmonod (Conservatoire et Jardin Botaniques de la Ville de Genève); Ana Juan (Universidad de Alicante); Stephen Jury (Reading University, UK); Costas Kadis* (Research Promotion Foundation, Cyprus); Georgia Kamari* (University of Patras); Wolf-Henning Kusber (Berlin Botanic Garden); Zacharias Kypriotakis* (Technological Education Institute, Kriti); Emilio Laguna Lumbreras* (Generalitat Valenciana); Antonino La Mantia (Università degli Studi di Palermo); Edwin Lanfranco* (University of Malta); Sandro Lanfranco; Leonardo Llorens Garcia* (Universitat de les Illes Balears); Pietro Lo Cascio; Christodoulos Makris; Jean-Francois Marzocchi; Joan Mayol* (Conselleria de Medi Ambient. Govern Balear); Frédéric Medail* (Université de Marseille III); Henri Michaud (Conservatoire Botanique National Méditerranéen); Juan Carlos Moreno Saiz (Universidad Autónoma de Madrid); Juan Francisco Mota Poveda (Universidad de Almería); Maurici Mus* (Universitat de les Illes Balears); Toni Nikolic* (University of Zagreb); Guilhan

Paradis*; Salvatore Pasta; Pietro Pavone (Fitogeografia della Sicilia, Università di Catania); Claudia Perini (Università di Siena); Dinitrios Phitos; Angélique Quilichini* (Université Paul Sabatier, Toulouse); Francesco Maria Raimondo* (Università di Palermo); Dominique Richard (European Topic Centre); Juan Rita Larrucea* (Universitat de les Illes Balears); Federico Selvi (Università degli Studi di Firenze); Darrin Stevens* (Malta Environment and Planning Authority); Angelo Troìa* (Riserva Naturale Orientata "Saline di Trapani e Paceco"); Nicholas J. Turland (Missouri Botanical Garden); Dimitris Tzanoudakis* (University of Patras); Paolo Usai* (Università degli Studi di Cagliari); Giuseppe Venturella* (Università di Palermo); Josep Vicens Fandos* (Universidad de Barcelona); Deryck E. Viney* (Herbarium of Northern Cyprus). The Royal Botanic Garden Edinburgh and the Botanischer Garten und Botanisches Museum Berlin-Dahlin, FU Berlin also kindly supplied herbarium specimen photos of plants for this work.

Proof-reading of the English version by Hazel Sharman and Patty Jacobs of Fairchild Tropical Botanic Garden, SSC volunteer William John Rogers and SSC communication staff Anna Knee, Andrew McMullin and Bryan Hugill (who also helped on the website) is gratefully acknowledged. Technical input by SSC Red List Programme Officer Craig Hilton-Taylor and Programme Officer Mariano Gimenez Dixon, who helped with the maps, was also much appreciated. Travis Gobeil provided the first design of this booklet which he unfortunately could not complete due to illness; we wish him the best for the future. The IUCN French National Committee has provided support on several occasions to the development of this project, for which they are greatly thanked.

Finally special thanks to Denis Landenbergue for his support throughout this project.

# *Preface*

This booklet presents a selection of 50 of the most threatened plant species growing on Mediterranean islands. It aims to draw the attention of both the public and politicians to the vulnerability of island floras in the Mediterranean, and calls for urgent conservation measures. The impact of increasing human activity and changes in agricultural practices must not lead to the extinction of these and other species.

As this booklet is aimed at the lay person, the text has been made as simple and non-technical as possible. Technical terms which were impossible to avoid are listed in the glossary. Readers who wish to learn more about the species highlighted in this book, or on plant conservation in general, can refer to the references at the end of this booklet or visit the 'Top 50' website (www.iucn.org/themes/ssc/plants/top50).

# Introduction

## Mediterranean Island plants

The Mediterranean basin contains nearly 5,000 islands and islets. While many of these are quite small (4,000 cover an area of less than 10 km$^2$), there are also many larger islands such as Sicily (with an area of 25,700 km$^2$). This great diversity in island size as well as differences in altitude and geology means that a large number of habitats are represented in the region. Different geographic situations (some islands are close, and others far away from the mainland), and their geographic history (some islands have been isolated for a long time, and others not), have produced a

Gulf of Valinco, Corsica.

flora of exceptional diversity. In addition, many Mediterranean plants are closely tied to traditional human activities which maintain this species richness.

Nearly 25,000 species of flowering plants and ferns are native to the countries surrounding the Mediterranean basin, and 60% of these plants are found nowhere else in the world. This extreme richness means that the Mediterranean is considered one of the world's 34 biodiversity "hotspots".

Thanks to their isolation on islands, some ancient plant species have managed to survive while their relatives on the mainland became extinct. This is because some mainland species could not compete with the migration into their habitat of new species, mainly caused by climate change during the last glacial periods. Because natural exchange of genetic material between the island and mainland species has been limited or non-existent, successive mutations caused the gradual formation of new species unique to each island.

The number of "endemic" species, that is those which are only found on one or a group of islands, is therefore very high. On the larger islands, around 10% of the species are endemic or unique to the island.

These endemic species are often very localized and have a small number of individuals, which makes them particularly susceptible to extinction. The 'Top 50' presented in this booklet have been selected from these many rare and threatened species.

An invasive species (*Carpobrotus edulis*), Porto Pollo, Corsica.

1

## The IUCN Red List Categories and Criteria

To evaluate whether or not a species is threatened, scientists measure the likelihood of extinction for each species. To do this they note whether or not the species is declining, either in number of individuals, or if the area in which it is found is getting smaller. They also take into account extreme population fluctuations because if the number of individuals is very small, a sudden change could drive the species to extinction.

The Species Survival Commission has developed the IUCN Red List Categories and Criteria, designed so that the threat level for all living species may be measured using precise, quantifiable criteria. The nine Red List Categories are as follows:

- ➤ Extinct (EX)
- ➤ Extinct in the Wild (EW)
- ➤ Critically Endangered (CR)
- ➤ Endangered (EN)
- ➤ Vulnerable (VU)
- ➤ Near Threatened (NT)
- ➤ Least Concern (LC)
- ➤ Data Deficient (DD)
- ➤ Not Evaluated (NE)

Samaria Gorge, Crete.

To chose to which Red List category a species belongs, a series of criteria have been developed, and these have been listed for each species in the "Why is it threatened?" section. These criteria range from A to E, with subcriteria from 1-4, a-c, and i-v. A species must meet one or more of these criteria to qualify for the category. As it is too complicated to explain how the criteria are used in this short introduction, readers who are interested are advised to read the IUCN Red List Categories and Criteria (ver. 3.1) available from IUCN or on the web at www.iucn.org/themes/ssc/red-lists.htm

Vegetation in front of the Minajdra Temple, Malta.

The list of the threatened species of a country or a region is called a Red List. The Red List measures the "state of heath" of biodiversity, noting why species are threatened and what conservation actions are needed to improve their situation. Red List assessments are regularly reviewed in order to monitor if a species' status is getting better or worse, and what new action may be needed to reduce the risk of extinction.

Assessment of the conservation status of the plants of the Mediterranean Islands is an important first step in planning conservation activities and in ensuring sustainable use of these species.

## Criteria for selecting the 'Top 50' Mediterranean Island species

The 50 species presented in this booklet were selected according to a number of criteria. These include level of threat, regional representation, an agreed taxonomy, and representation of different plant families and life forms. A mushroom species, which technically is classified as a fungus and not a plant, has been included to further represent the diversity occurring in the Mediterranean.

Sadly, many more species than those listed in this booklet require urgent conservation measures. The IUCN/SSC Mediterranean Island Plant Specialist Group, in collaboration with scientists and Red List activities taking place in all the countries in the region, is continuing to identify threatened species and propose conservation actions to ensure that the conservation status of the unique and beautiful flora of the Mediterranean Islands improves. Additional species sheets will be added to the Group's website gradually (www.iucn.org/themes/ssc/sgs/mipsg) and the status of species already listed in the booklet will be monitored.

The species included in this booklet are listed in the table of contents. They are found on 12 islands or island groups. While most (46) are classified as Critically Endangered (CR), examples of species evaluated as Extinct in the Wild (EW), Endangered (EN) and Data Deficient (DD) are also included.

## Threats to the Mediterranean flora

The main factor raising the risk of extinction for the 50 species listed in this booklet is linked to the size of their population and their distribution. In almost every case, due to the small number of individuals or tiny area of distribution, any major disturbance (for example, fire or construction work) might just push the species to extinction or seriously reduce its chances of survival. For four of the species listed, fewer than 50 individuals are known in the wild.

In addition, when the number of individuals in a population falls below a certain threshold, the species loses genetic diversity which reduces its ability to adapt to change, and therefore further increases its extinction risk.

The main threats faced by the species evaluated in this booklet, and by extrapolation, for numerous other Mediterranean plant species, are mostly due to direct or indirect human activities. These fall under the following categories (in decreasing order of importance):

➤ Urbanization

➤ Tourism and recreation

➤ Fires

➤ Change in agricultural practices (intensification or abandonment)

➤ Invasive alien species

➤ Collecting pressure

It is also clear that climate change will increase these threats. In effect, not only are plants by their nature relatively immobile, it is also difficult to

Gulf of Propriano, a locality for *Anchusa crispa*, Corsica.

Sheep grazing on the island of Crete.

change altitude if the species relies on specific ecological conditions. Island floras also have limited scope to migrate horizontally, especially on small islands.

## Recommendations for conservation action

Nearly three quarters of the 50 species selected for this booklet benefit from some sort of legal protection, whether at a national or international level. About half have some or all of their population included in a pro-tected area. However these conservation measures, while very valuable, are often not sufficient to completely reduce the risk of extinction, notably due to problems in applying the law as well as inadequate manage-ment of the protected areas.

Half of the species listed are conserved in botanical gardens or seedbanks (known as *ex situ* conservation or cultivation), but there is no guarantee that if the species disappears in the wild, it will be possible to reintroduce the species or to maintain its genetic diversity in the long-term. It is in all cases preferable, and certainly less costly, to try to maintain the species in its natural habitat (known as *in situ* conservation). However, when a species does become threatened, cultivation can serve as an "insurance policy" in case the species becomes extinct in the wild. A good example of this is *Diplotaxis* from the island of Alborán. The species disappeared from the island but has been re-introduced using cultivated plants.

Conservation for each species requires specific, targeted actions. These are outlined in each datasheet and fall under the following categories (in decreasing order of importance for the 50 species studied):

➤ Legal protection at regional, national or international levels

➤ Improved biological and ecological knowledge to better target conservation measures

➤ Establishment of management plans for the species and its habitat

➤ Creation and management of protected areas

➤ Cultivation in botanical gardens or main-tenance in seedbanks

➤ Reintroduction or reinforcement of popula-tions

➤ Management of grazing

➤ Control of invasive species

➤ Fire prevention

Blue Grotto, Malta. An example of friable seacliffs rich in threatened species.

The planning, financing and implementation of conservation measures require substantial resources that can only be put into place by decision-makers, managers and the public who are convinced by their importance and relevance to people. It is therefore essential to increase public awareness about the importance of plant conservation and its fundamental value to human well-being.

Protection of the shore above high water level, Ajaccio, Corsica.

## Structure of the species datasheets

The 50 species are presented in alphabetical order of the islands or groups of islands on which they are found, and within that group, in alphabetical order by Latin name. Because some species have many common names and others none, the species have been listed according to their Latin name. In some cases when the species was renamed after its taxonomic status changed, a Latin synonym has been listed.

Each species sheet includes the following sections:

➤ The Latin name as well as any local common name which exists

➤ A photo or drawing of the species (in three cases the species is so rare and poorly-known that no good photo exists)

➤ A small map showing where in the Mediterranean it is found

➤ General distribution information. The exact location where these species occur is not given as readers are asked not to try to find the species until their conservation status has improved

➤ A brief description

➤ Additional information and interesting facts

➤ Threat category and reasons for it

➤ Existing conservation measures both *in situ* and *ex situ*

➤ Proposed conservation actions

➤ Main contributors to the datasheet

Specific references for each datasheet have been left out of the booklet due to space limitations, but all relevant references are included in the Top 50 website.

Bertrand de Montmollin
Chair, Mediterranean Island Plant Specialist Group
Rue de la Serre 5, CH-2000 Neuchâtel, Switzerland
gspim@biolconseils.ch

Wendy Strahm
Plants Officer, Species Programme
IUCN – The World Conservation Union
Rue Mauverney 28, CH-1196 Gland, Switzerland
was@iucn.org

# Silene hicesiae

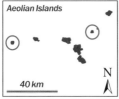

Aeolian Islands

40 km

N

PIETRO LO CASCIO

ANGELO TROIA

| | |
|---|---|
| **Latin name:** | *Silene hicesiae* Brullo & Signorello |
| **Common name:** | **Silene vellutata delle Eolie (Italian)** |
| **Family:** | **Caryophyllaceae (carnation family)** |
| **Status:** | **CRITICALLY ENDANGERED (CR)** |

### Where is it found?

This species is endemic to the Aeolian Islands, and grows on the rocky slopes of two small volcanic islets: Panarea (where the population covers an area of 3-4 hectares with almost 400 individuals) and Alicudi (with an even smaller population of less than 30 individuals covering 60 m²). These two populations are situated about 60 km apart.

### How to recognise it

This perennial plant has a woody base and grows to between 50-120 cm tall. It produces both sterile and fertile rosettes of densely hairy, elliptical leaves between 5-10 cm long. The fertile rosettes produce a hairy, usually unbranched flowering stem with bunches of five-petalled pink flowers which open in May. The fruits mature between the second half of August and the beginning of September.

### Interesting facts

*Silene hicesiae* belongs to the "*Silene mollissima* group", which comprises seven species endemic to the coastal cliffs of the western Mediterranean basin. These different species may have evolved when the Mediterranean became drier during the late Miocene period, about 5 million years ago, and the ancestral species became isolated into several different populations. This is one of numerous examples demonstrating how geographic separation of populations belonging to the same species, for example on islands, can contribute to the formation of new species.

## Why is it threatened?

This species is categorized CR (Critically Endangered) according to IUCN Red List Criteria B1ab(iv,v)+2ab(iv,v). This means that the species covers a very small area, the populations are severely fragmented, and the number of populations and mature individuals is declining. The population of the islet of Alicudi comprises less than 50 mature individuals, so few that its future on this islet is in jeopardy unless urgent conservation measures are taken.

The main threats that this species faces include wildfires; grazing by herbivores (e.g. rabbits); invasive alien plants (e.g. "Tree of Heaven" *Ailanthus altissima*); and incorrect management of protected areas. Any of these threats could wipe out either of the two populations.

## What is being done to protect it?

**Legally:** This species is listed as a priority species in Annexes II and IV of the EC Habitats Directive. The islets of Alicudi and Panarea are nature reserves where the collection of *Silene hicesiae* is strictly forbidden.

*In situ & ex situ:* A number of conservation measures have been implemented as part of the EU LIFE project "Conservation of priority species of the Eolian Islands' flora – EOLIFE99": improvement of knowledge on the biology and ecology of *Silene hicesiae*, reinforcement of wild populations with plants propagated *ex situ*, cultivation in botanical gardens and seedbanks, and informing decision makers and the general public. This plant is also included in the GENMEDOC project (an inter-regional network of Mediterranean seedbanks), and seeds are being collected in order to propagate this species.

## What conservation actions are needed?

All populations of the "Tree of Heaven" *Ailanthus altissima* near to where *Silene hicesiae* grows need to be eliminated. The conservation programme initiated by EOLIFE99 also needs to be continued.

## Scientific coordination

Dr Angelo Troìa, Regional Nature Reserve "Saline di Trapani e Paceco", WWF-Italia, Trapani, Italy. Dr Salvatore Pasta, freelance botanist, Palermo, Italy.

# *Diplotaxis siettiana*

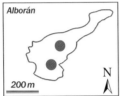

| | |
|---|---|
| Latin name: | *Diplotaxis siettiana* Maire |
| Common name: | Jaramago de Alborán (Spanish) |
| Family: | Cruciferae (mustard family) |
| Status: | CRITICALLY ENDANGERED (CR) |

### Where is it found?

Endemic to the Spanish island of Alborán, this species was last seen in 1974, when seeds were fortunately collected before the species disappeared from the island. A re-introduction in 1999 appears successful, although given extreme population fluctuations each year, more time is needed to ensure that the re-introduced population is self-sustaining.

Alborán is the top of a volcanic platform situated between Spain and Morocco, around 50 km from the nearest continent. This small island (600 m x 200 m) resembles an aircraft carrier due to its flat surface, reaching 10 m above sea level and surrounded by steep, almost vertical cliffs. The island has a lighthouse and is now used as a military base.

In 1974 the plant was found growing in a tiny area around the helicopter platform. The weedy nature of *Diplotaxis siettiana* makes it fairly tolerant to human disturbance. In fact it does not grow in the more stable vegetation dominated by *Frankenia pulverulenta* and *Mesembryanthemum nodiflorum*, both apparently tolerant to high concentrations of salt and/or nitrogen. Rainfall is very low, with less than 100 mm per year.

### How to recognise it

This annual spreading herb is 10-40 cm tall and has sparse hairs. The deeply-lobed leaves are somewhat fleshy, 5-15 cm long, and initially form a rosette at the base of the plant. Along the stem the leaves are smaller and the divisions narrower. With sufficient rain between December and April, the plant produces numerous yellow, 4-petalled flowers. The fruits contain many seeds and are up to 3 cm long, ripening between February and May.

## Interesting facts

This species is the only representative of the mustard family in the poor flora of Alborán (10 species). The island has long been inhabited, first by lighthouse keepers and later by the military which built the helicopter platform, widened the former small harbour, and constructed some temporary dwellings near the lighthouse.

## Why is it threatened?

This species has been categorized CR (Critically Endangered) according to IUCN Red List Criteria B1ac(iv)+2ac(iv). This means that the plant grows in a very small area and population numbers fluctuate greatly. Irrigation of the area with sea water where the plant was originally found, in order to reduce dust for helicopter landings, was the direct cause of extinction of this species. The island's fragile habitat has been largely modified by humans who recently introduced some domestic animals, causing further soil erosion and nitrification. Germination, flowering and fruiting are dependent on rainfall. An observation in 1970 referred to possibly hundreds of adult individuals, although only 150 were recorded in 1974, and none afterwards. In 1999, 48 plants were re-introduced but scientists are not sure that the population is self-sustaining. No dispersal to other parts of the island has taken place

## What is being done to protect it?

**Legally:** This species is protected at the regional and European level. It is listed as a priority species in Annexes II and IV of the EC Habitats Directive.

*In situ*: The islet has been declared a Maritime-Terrestrial Reserve and Natural Landscape ("Paraje Natural") by the Andalusian government, and nominated for the European Natura 2000 Network. Various re-introductions in 1988 and 1996 were apparently unsuccessful. Another in 1999 is being supplemented with reinforcement and monitoring campaigns.

*Ex situ*: Fortunately before its extinction, seed had been collected and multiplied at the seedbank of the Escuela T.S. de Ingenieros Agrónomos (Universidad Politécnica, Madrid), and distributed to some botanical gardens. In cultivation high germination rates can be achieved.

## What conservation actions are needed?

This species should be added to the Spanish National Catalogue of Threatened Species, listed in the highest category (Endangered). Ideally the island should be designated as a strict Nature Reserve. If this is impossible, habitat restoration and on-going management still needs to be carried out indefinitely, including alien species eradication, monitoring, and no new construction of infrastructure. Periodic reinforcement campaigns in order to maintain the population might be necessary. The fact that *Diplotaxis siettiana* seems to prefer semi-disturbed habitat and competes poorly with dominant species must always be taken into account.

## Scientific coordination

Dr Juan Carlos Moreno Saiz, Universidad Autónoma de Madrid, Spain.
Professor Juan Francisco Mota Poveda, Universidad de Almería, Spain.
Professor César Gómez-Campo, Universidad Politécnica de Madrid, Spain.

# Apium bermejoi

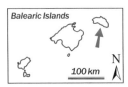

| Latin name: | *Apium bermejoi* L. Llorens |
|---|---|
| Common name: | Apid den Bermejo (Catalan) |
| Family: | Umbelliferae (celery family) |
| Status: | CRITICALLY ENDANGERED (CR) |

### Where is it found?

This species is endemic to the Balearic Islands, and occurs in the eastern part of the island of Minorca. Here it is only found in two small areas separated by a rocky zone about 200 m wide. The total population numbers less than 100 individuals and covers an area of just a few dozen square metres. *Apium bermejoi* grows in stream beds that dry out in summer and occasionally during dry winters. It grows on acidic soil which accumulates in small rock depressions, and requires only moderate sun exposure.

### How to recognise it

This herbaceous plant creeps over the ground. Its hollow stem is equipped with narrow grooves that secrete aromatic oils, giving the plant a celery smell. Its leaves are compound, with about 10 toothed leaflets. The flowers are white and arranged in small umbrella-shaped inflorescences. *Apium bermejoi* usually flowers between April and May, but may flower at other times of the year depending on the weather.

### Interesting facts

This perennial plant reproduces from seeds but can also reproduce vegetatively from stolons, which are horizontal stems creeping just above the ground having the capacity to take root and form new plants. There is evidence that the two populations on either side of the rocky zone are genetically different, but the importance of this for the long-term maintenance of the entire population has yet to be assessed. This species benefits from nitrogen provided in seabird droppings.

## Why is it threatened?

This species has been categorized CR (Critically Endangered) according to IUCN Red List Criteria B1ab(v)c(iv)+2ab(v)c(iv); C2a(i); D, meaning that it only occurs in a single site and the total number of individuals is very small and fluctuates in number. The most recent census counted 98 individuals, although many were young plants which never reached reproductive age. In other years the total population numbered less than 60 individuals, again not all reaching maturity. Given the threats facing this species, a continued decline in number of individuals is predicted.

The species is threatened both directly and indirectly. Its habitat is extremely unstable with available water and nutrients varying greatly from year to year. *Apium bermejoi* does not support competition from other species very well, including competition from native carpet-forming species as well as introduced alien species such as *Carpobrotus edulis*. It is directly threatened by trampling by fishermen and hikers, as well as from motorbikes on the beach.

Climate change may cause changes in its habitat. For example, several consecutive dry years will weaken this species and favour the development of opportunistic, more competitive species. Any wild collection of *Apium bermejoi* represents a potential threat.

## What is being done to protect it?

**Legally:** This species is listed in Annex I of the Spanish Royal Decree 439/1990 which grants it protection in its natural site. Internationally, *Apium bermejoi* is included in two legal documents: Appendix I of the Bern Convention and Annexes II and IV of the EC Habitats Directive, where it is listed as a priority species.

*In situ*: Since 2003, the University of the Balearic Islands has started a rehabilitation programme for *Apium bermejoi* in its natural habitat with financial support from the MAVA Foundation. Since 1996, a programme to eradicate all species of the introduced and invasive *Carpobrotus* in natural areas of the Balearic Islands has progressed with mixed success. Since 2002, the eradication programme has focused on sites with rare species such as *Apium bermejoi*.

*Ex situ*: Seeds of this species are stored in the Sóller Botanical Garden seedbank, located on Majorca, where the plant is also under cultivation.

## What conservation actions are needed?

It is essential that the small area where this species grows be protected from trampling and motorbikes. To make conservation actions more effective, studies on population dynamics (recruitment and mortality) and reproductive biology of this species are needed. (Re-)introduction of this species to suitable habitats to increase its number of populations and its survival chances is needed.

## Scientific coordination

Dr Maurici Mus, Dpto. Biologia, Universitat de les Illes Balears, Palma de Mallorca, Spain.
Dr Juan Rita Larrucea, Dpto. Biologia, Universitat de les Illes Balears, Palma de Mallorca, Spain.

# Arenaria bolosii

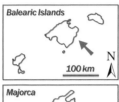

| | |
|---|---|
| Latin name: | *Arenaria bolosii* (Cañig.) L. Sáez & Rosselló |
| Common name: | none |
| Family: | Caryophyllaceae (carnation family) |
| Status: | CRITICALLY ENDANGERED (CR) |

### Where is it found?

This species is only known from a single site on the island of Majorca in the central part of the Tramuntana mountain range. Less than 200 individuals grow in an area covering approximately one hectare. The species is very sensitive to competition from other species and therefore grows in open areas with little soil cover. It is found above an altitude of 900 m on north-facing slopes.

### How to recognise it

This herbaceous perennial grows in loose clumps, reaching a height of 5-10 cm. Several stems covered by short hairs branch out from its base. The leaves are arranged in pairs and are green, occasionally tinged with pink or grey. Its small white flowers have five petals and develop at the tips of the stems. This species flowers and fruits between June and July.

### Interesting facts

Little is known about the biology of this plant, and there is even disagreement over its taxonomic status. While *Flora Iberica* treats it as a subspecies (*Arenaria grandiflora* L. subsp. *bolosii* (Cañig.) Küpfer), others consider it a full species and it is listed in the

Spanish Red Book as such. The plant is present throughout the year, even in winter. Its small size and buds close to the ground protect it from the wind and make it more tolerant of dry conditions. Note that the closely-related species *Arenaria grandiflora* grows in the mountains of southern and central Europe where it has not been recorded as threatened.

### Why is it threatened?

*Arenaria bolosii* has been categorized CR (Critically Endangered) according to IUCN Red List Criteria B1ab(iii,v)+2ab(iii,v); C2a(ii). This means that the area where it is found and the number of individuals are very small, it is known from a single site, and the habitat where it grows as well as the number of mature individuals continues to decline.

Plant collectors seeking botanical rarities directly threaten this species. In addition, it seems to hybridize with a closely related species, *Arenaria grandiflora* subsp. *glabrescens*. Other threats include hikers trampling the plant, fires, and habitat modification as more people use the area for hiking and camping.

### What is being done to protect it?

**Legally:** This plant's habitat is protected by Law 1/1991 of the Parliament of the Balearic Islands as a Natural Site of Special Interest. It is listed as Critically Endangered in the Red List of the Spanish vascular flora (*Lista Roja de la flora vascular española*) as well as in the Spanish Red Book (*Atlas y Libro Rojo de la flora vascular amenazada de España*), although this does not confer any specific legal protection.

*In situ*: No measures taken yet.

*Ex situ*: Seeds of this plant are stored in the seedbank of the Botanical Garden of Sóller, but it is possible that these are hybrids with *Arenaria grandiflora* subsp. *glabrescens*.

### What conservation actions are needed?

This species merits increased legal protection, such as inclusion in the Annexes of the Bern Convention and the EC Habitats Directive. Public access to the site must be controlled. Urgent conservation measures must be taken which include reinforcing the number of plants that remain at this last site and reintroducing the species into other areas where it was previously known.

If the seeds stored in seedbanks are found to be hybrids, any re-introduction campaign using this material will be detrimental to conservation. Genetic studies are therefore needed, and seeds should be collected only when it has been established that they belong to this species.

### Scientific coordination

Dr Gabriel Bibiloni, Dpto. Biologia, Universitat de les Illes Balears, Palma de Mallorca, Spain.
Dr Maurici Mus, Dpto. Biologia, Universitat de les Illes Balears, Palma de Mallorca, Spain.

# *Brimeura duvigneaudii*

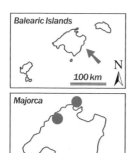

Balearic Islands

100 km

N

Majorca

40 km

N

| | |
|---|---|
| **Latin name:** | *Brimeura duvigneaudii* (L. Llorens) Rosselló, Mus & Mayol |
| **Common name:** | none |
| **Family:** | Hyacinthaceae (hyacinth family) |
| **Status:** | CRITICALLY ENDANGERED (CR) |

## Where is it found?

Endemic to Majorca, this species only occurs in very small numbers at three localities. It grows in limestone rock crevices and slopes near the sea at an altitude of approximately 150-250 m. It thrives in hot, sunny places.

## How to recognise it

This plant is small and inconspicuous, not exceeding 10 cm in height. Its grass-like, ribbed, linear leaves are about 3 mm wide and triangular in cross-section. The one-sided inflorescences are composed of two to five (rarely seven) flowers with pink corollas (never blue like those of its close relative *Brimeura amethystina*). The best way to tell this species apart from the closely related *Brimeura fastigiata* is that the lobes of the flower are shorter than its tube.

## Interesting facts

The above-ground parts of the plant die back in summer when it is hottest and driest. Only one population has so far been observed producing seeds; all others seem to reproduce asexually by bulb division. *Brimeura duvigneaudii* is considered to be a relict species, a remainder from a once larger group that, in the course of climatic change, has nearly disappeared over the millennia. These remnant populations may have difficulties coping with today's climate.

### Why is it threatened?

This species has been categorized CR (Critically Endangered) according to IUCN Red List Criteria B1ab(iv,v)+2ab(iv,v); C2a(i). This means that it only occurs in three fragmented localities, and that the populations are small, declining, and are estimated to have no more than 50 mature individuals each (although estimates are difficult given the plant's inconspicuous habit). One population has nearly disappeared. The area of occurrence of 7.5 km$^2$ is small enough to make this species vulnerable even to extreme natural events, such as fires. The species also faces recruitment problems, possibly due to climate.

### What is being done to protect it?

**Legally:** On a regional level, the habitat is protected as a Natural Site of Special Interest by law 1/1991 of the Parliament of the Balearic Islands. It is listed in the Red List of the Spanish vascular flora (*Lista Roja de la flora vascular española*) as well as in the Spanish Red Book (*Atlas y Libro Rojo de la flora vascular amenazada de España*), although this does not confer any specific legal protection. The species itself is listed in the catalogue of threatened species of the Balearic Islands (*Catálogo Nacional de especies amenazadas*, CNEA).

*In situ*: No measures taken yet.

*Ex situ*: No measures taken yet.

### What conservation actions are needed?

It is necessary to undertake a careful search for other populations of *Brimeura duvigneaudii*, which may be easily overlooked due to its minute size. A substantial increase in monitoring is needed to understand the population dynamics and reproductive behaviour of this species. Agricultural fires that are set to provide grazing for sheep should not be allowed in the area where *Brimeura* grows. Cultivation of this species in a botanical garden would be useful.

### Scientific coordination

Dr Gabriel Bibiloni, Dpto. Biologia, Universitat de les Illes Balears, Palma de Mallorca, Spain.
Dr Maurici Mus, Dpto. Biologia, Universitat de les Illes Balears, Palma de Mallorca, Spain.

# *Euphorbia margalidiana*

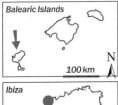

Balearic Islands

100 km

N

Ibiza

20 km

N

| | |
|---|---|
| **Latin name:** | *Euphorbia margalidiana* Kühbier & Lewej |
| **Common name:** | Lletrera (Catalan) |
| **Family:** | Euphorbiaceae (spurge family) |
| **Status:** | CRITICALLY ENDANGERED (CR) |

### Where is it found?

This species is known only from a single site, where some 200 individuals grow along the rocky shore of the islet of Ses Margalides, close to the north-western coast of Ibiza. *Euphorbia margalidiana* grows in the cracks of friable limestone cliffs and boulders.

## How to recognise it

This shrub or small tree has succulent stems which are swollen with water, allowing the plant to tolerate periods of extended drought. Its leaves are light bluish-green, and mostly drop off at the end of spring, growing again in autumn, which is an adaptation to summer droughts. The flowers appear between March and April and are arranged in umbrella-like inflorescences. The fruits consist of three valves fused together which burst open when dry, projecting tiny seeds far away from the parent plant from June to July.

## Interesting facts

*Euphorbia margalidiana* is a perennial plant that requires good light and high temperatures. Two other species of *Euphorbia* are endemic to the Balearic Islands: *Euphorbia maresii* (which is divided into two subspecies, *maresii* and *balearica*) is not threatened; however *Euphorbia fontqueriana* is Critically Endangered and only found on the island of Majorca.

## Why is it threatened?

This species has been categorized CR (Critically Endangered) according to IUCN Red List Criteria B1ab(v)+2ab(v). This means that the plant is only known from one small locality and the number of individuals is declining.

The unique population covers a very small area of about eight hectares and consists of no more than 200 individuals. An increasingly dry environment and the risk of collapse of the cliffs where this plant occurs also present a threat. Monitoring of this species is difficult because the cliffs are unstable and dangerous.

## What is being done to protect it?

**Legally:** This species is listed in Annex II of the Ministerial Decree 22112 (1984) as a species of special national interest to be protected in the Balearic Islands. It is illegal to undertake any activity that could damage this plant. *Euphorbia margalidiana* is listed in Annex I (in danger of extinction) of Decree 439/90, which guarantees it protection in its native habitat. It is also listed in the Spanish Red Book (*Atlas y Libro Rojo de la flora vascular amenazada de España*). Internationally, it is included in Appendix I of the Bern Convention and as a priority species in Annexes II and IV of the EC Habitats Directive.

*In situ*: No measures taken yet.

*Ex situ*: This species is in cultivation in several places including the botanical gardens of Sóller (Majorca) and Marimurtra (Barcelona) in Spain. Seeds are also being conserved in seedbanks. Currently studies on the genetic variability of material held *ex situ* are being undertaken.

## What conservation actions are needed?

Access to the site where *Euphorbia margalidiana* grows should be prohibited. The Government of the Balearic Islands is currently conducting a feasibility study to introduce this species to another islet, but to date there is insufficient data to make recommendations for benign introductions to other sites.

## Scientific coordination

Dr Josep Vicens Fandos, Universidad de Barcelona, Laboratori de Botanica, Facultat de Farmacia, Universitat Barcelona, Spain.
Dr Maurici Mus, Dpto. Biologia, Universitat de les Illes Balears, Palma de Mallorca, Spain.

# *Femeniasia balearica*

**Balearic Islands**

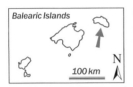

100 km

N

**Minorca**

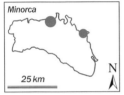

25 km

N

| | |
|---|---|
| Latin name: | *Femeniasia balearica* (J.J. Rodr.) Susanna |
| Synonym: | *Centaurea balearica* J.J. Rodr. |
| Common name: | **Soccarrell bord (Catalan)** |
| Family: | **Compositae (daisy family)** |
| Status: | CRITICALLY ENDANGERED (CR) |

### Where is it found?

Endemic to the Balearic Islands, *Femeniasia balearica* now occurs only in six sites in the north of the island of Minorca. It is a typical seashore species that grows in dry sunny places on sandy soils.

### How to recognise it

*Femeniasia balearica* is a small shrub or tree about 150 cm in height, covered by numerous sharp spines about 1 cm long, arranged in groups of three. It has two different types of leaves; those produced in spring are linear and entire, whereas the summer leaves are divided. The flowers are gathered together in yellow heads, surrounded by spines, and open between May and July, with fruits ripening at the end of summer and beginning of autumn.

### Interesting facts

This woody perennial reproduces from seed. Its spines protect it from herbivores and trampling, but unfortunately not from motor vehicles. In some specialist literature the genus *Femeniasia* is considered to be synonymous with *Centaurea*. This means that some lists of protected species cite this plant under the name of

*Centaurea balearica*, demonstrating both the practical as well as legal problems that changes of nomenclature can cause.

## Why is it threatened?

This species has been categorized CR (Critically Endangered) according to IUCN Red List Criteria B1ab(iii,v)+2ab(iii,v). This means that the area where this species is found covers less than 100 km², the extent and/or quality of habitat is declining, its total population is extremely fragmented, and number of individuals is declining.

    *Femeniasia balearica* covers a very small area and fewer than 2,200 mature individuals are found in six subpopulations. It is threatened by building and road construction, although these activities may also create opportunities for colonization. This species is often removed from beach paths because of its spines. Several individuals disappeared when a land owner planted pines *(Pinus halepensis)* in one of its subpopulations. It has been observed that in some years the activities of wood-eating beetles (*Oxythrea funesta* and *Tropinota hirta*) seem to reduce seed germination.

## What is being done to protect it?

**Legally:** This species is listed in Annex I (in danger of extinction) of the Spanish Royal Decree 439/1990 that guarantees it protection in its natural habitat. Internationally, *Femeniasia balearica* is included in Appendix I of the Bern Convention and as a priority species in Annexes II and IV of the EC Habitats Directive (in both cases under the name *Centaurea balearica*). Under the Birds Directive, the European Union protects the habitat of this species as an Important Bird Area.

*In situ*: A re-establishment plan for *Femeniasia balearica* is currently being drafted at the University of the Balearic Islands.

*Ex situ*: Seeds are being stored and individuals are cultivated at the Botanical Garden of Sóller (Majorca).

## What conservation actions are needed?

Research is needed to understand this species' population dynamics (births and deaths per year or per generation) and the effect of competition and herbivory on young plant survival. The re-establishment plan that is being drafted needs to be completed and implemented. It is extremely important that motorized vehicles be prohibited from driving in the area containing this species.

## Scientific coordination

Dr Maurici Mus, Dpto. Biologia, Universitat de les Illes Balears, Palma de Mallorca, Spain.
Dr Juan Rita Larrucea, Dpto. Biologia, Universitat de les Illes Balears, Palma de Mallorca, Spain.

# *Ligusticum huteri*

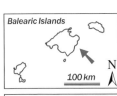

Balearic Islands

100 km

N

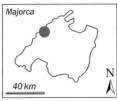

Majorca

40 km

N

HERBARI VIRTUAL DE LES ILLES BALEARS, http://herbarivirtual.uib.es

| | |
|---|---|
| Latin name: | *Ligusticum huteri* Porta & Rigo |
| Synonym: | *Coritospermum huteri* (Porta) L. Sáez & Rosselló |
| Common name: | none |
| Family: | Umbelliferae (celery family) |
| Status: | CRITICALLY ENDANGERED (CR) |

## Where is it found?

Endemic to the Balearic Islands, this species occurs in northern Majorca in the Tramuntana mountain range at 1,300-1,400 m in altitude. *Ligusticum huteri* grows on limestone cliffs in shaded, slightly moist rock crevices and on rocky ridges. No more than 100 individuals exist, distributed over an area less than 0.5 km$^2$.

## How to recognise it

*Ligusticum huteri* is a perennial plant growing up to 100 cm in height, which forms tufts with sprawling roots and has a peculiar smell. Its stem is stiff and grooved, and shiny, deeply incised leaves grow from the base of the plant. The flowering period occurs from June to August, when large, dense bunches of white flowers in umbrella-shaped inflorescences appear. The flower petals have a dark grey line on their lower side. The seeds are furrowed and flattened sideways.

### Interesting facts

The biology of this plant is not well known. The parts above ground disappear during winter and grow up from the base between April and June. The plant only flowers abundantly after several years of favourable conditions. This initial period is probably necessary for the accumulation of energy reserves. *Ligusticum huteri* reproduces from seed only.

### Why is it threatened?

This species has been categorized CR (Critically Endangered) according to IUCN Red List Criteria B1ab(v)+2ab(v); C2a(ii). This means that it only occurs in one locality over a very small area (0.5 km²), and that the total number of individuals is small and in decline. At least 50% of the population has disappeared over the past 10 years due to drought and increased grazing pressure from wild goats. Bush fires started by sheep breeders also pose a threat.

### What is being done to protect it?

**Legally:** At a regional level, *Ligusticum huteri* is included as a species sensitive to habitat changes in the catalogue of threatened species of the Balearic Islands (*Catálogo Nacional de especies amenazadas*, CNEA), which grants it protection in its natural habitat. Its habitat is protected as a Natural Site of Special Interest by Law 1/1991 of the Parliament of the Balearic Islands, and the species occurs in a military zone with limited access. At a national level this species is listed in the Spanish Red Book (*Atlas y Libro Rojo de la flora vascular amenazada de España*), and at an international level the site is listed as an Important Bird Area in the EU Birds Directive.

*In situ*: In 1998 a five-year plan was put into action by the Botanical Garden of Sóller to re-establish the plant in its natural habitat, of which some aspects are still in development. The Government of the Balearic Islands protects part of its population with fencing.

*Ex situ*: Seeds are stored in a seedbank at the Botanical Garden of Sóller and some plants are maintained in culture.

### What conservation actions are needed?

It is imperative to control the goat population on the site and ensure that there is no botanical collection. Sheep breeders should be prevented from setting bush fires. Possibilities of introducing this species to other mountainous regions of the island should be investigated.

### Scientific coordination

Mr Joan Mayol, Conselleria de Medi Ambient. Govern Balear, Palma de Mallorca, Spain.
Dr Gabriel Bibiloni, Dpto. Biologia, Universitat de les Illes Balears, Palma de Mallorca, Spain.

# Lysimachia minoricensis

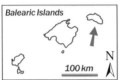

Balearic Islands

100 km

N

Minorca

25 km

N

| | |
|---|---|
| Latin name: | *Lysimachia minoricensis* J.J. Rodr. |
| Common name: | none |
| Family: | Primulaceae (primrose family) |
| Status: | EXTINCT IN THE WILD (EW) |

### Where is it found?

Endemic to Minorca in the Balearic Islands, *Lysimachia minoricensis* was only known from a single location (Barranc de Sa Vall), where it disappeared between 1926 and 1950. Fortunately seeds had been collected, and the species was cultivated from 1926 in the Botanical Garden of Barcelona. Although *Lysimachia minoricensis* was believed to be lost when the garden was abandoned during the Civil War, a colony was later rediscovered, growing in the shelter of a bushy thicket. The only notes made by its discoverer recorded that the species grew in cool, shady places.

### How to recognise it

This herbaceous biennial species varies between 25-80 cm in height. Its stem is upright, simple or branching from the base, with small glands in its upper parts. Its oval green leaves have almost no or a very short stalk, and are covered by whitish nerves on the upper side and purple beneath, a feature that is common in some other Balearic plants. The small flowers are arranged in a loose, terminal bunch with leafy bracts. They are yellowish-green with a red-violet throat, and are 4 mm long (just a little longer than the calyx). The calyx is deeply divided with obtuse teeth. Flowering from May to July, its fruits, 3.5-5 mm long, contain numerous black, 1 mm long, rough, laterally compressed seeds.

### Interesting facts

This species seems able to produce seeds without pollinators. The number of seeds produced per individual is very high, reaching up to 3,300. Experiments have shown that germination rates are very high, and germination can occur over a wide range of temperature, light, and soil salinity conditions. The leaves emit a strong odour that may be an adaptation to protect the plant from herbivores.

### Why is it threatened?

This species has been categorized EW (Extinct in the Wild) according to the IUCN Red List Criteria. This means that the species is now only found in cultivation and seedbanks. Note that while the species has recently been re-introduced to the wild, it has not yet formed self-sustaining populations. The reasons for its disappearance in the wild are unknown. It is possible that over-collection and the impact of human activities (such as fire and changes in agricultural practices) may have caused its extinction. On the other hand, it is possible that this species might have benefited from agricultural activities practised in the past, and that the cessation of these practices may have caused the disappearance of habitat favourable to this species. The most successful re-introduction attempts, where plants survived for up to five years, were in areas previously disturbed by fire, cattle or goats.

### What is being done to protect it?

**Legally:** This species is included in Appendix I of the Bern Convention and listed in the national catalogue of threatened species (*Catálogo Nacional de especies amenazadas*, CNEA). The natural area where it was known and where re-introduction attempts have been made (Son Bou i Barranc de Sa Vall) is designated as a Site of Special Natural Interest by the Law 1/1991 of the Parliament of the Balearic Islands. It is also included in the European Natura 2000 network.

*In situ*: Attempts to re-introduce the species into its native habitat have been undertaken since 1959 but have been unsuccessful. The most recent attempts have re-introduced this species with mycorrhizal fungi in the gorges of Sa Vall, Trebaluger, and Algendar. However, the seeds from these individuals have failed to germinate, so this species is still considered to be Extinct in the Wild.

*Ex situ*: Seeds of this species are conserved in numerous seedbanks. It is also cultivated in several botanical gardens using seeds produced by the original specimens.

### What conservation actions are needed?

Most urgently, it is important to understand the reproductive biology of this species, especially the factors that inhibit the germination of seeds in the wild. Second, re-introduction attempts need to be continued. Finally a management plan needs to be developed for the areas where the species has been re-introduced so that the re-introduced populations become self-sustaining.

### Scientific coordination

Dr David Galicia Herbada, TRAGSA, Área de Medio Ambiente, Madrid, Spain.
Mr Pere Fraga Arguimbau, Consell Insular de Menorca, Spain.

# *Naufraga balearica*

**Balearic Islands**

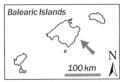

100 km

N

**Majorca**

40 km

N

| | |
|---|---|
| **Latin name:** | *Naufraga balearica* Constance & Cannon |
| **Common name:** | none |
| **Family:** | Umbelliferae (celery family) |
| **Status:** | CRITICALLY ENDANGERED (CR) |

## Where is it found?

*Naufraga balearica* occurs only on the island of Majorca in the Balearic Islands, where it is found at the base of coastal cliffs to the north of the island. It usually grows in shady cracks in boulders where a little calcareous soil or clay has accumulated, and needs humid conditions. Another population of what is believed to be the same species was discovered in 1981 on the western coast of Corsica, between Cargèse and Piana, but this had disappeared by 1983.

## How to recognise it

This perennial herbaceous plant is recognised by its very small size for a member of the celery family (only 2-4 cm tall), its numerous lateral shoots which emerge from the base of the plant in spring, and a simple rather than compound umbrella-shaped inflorescence. Its leaves are grouped in a rosette, each one divided into three to five oval leaflets which lie flat on the ground. The pink flowers are very small (about 2 mm in diameter) and open between June and July.

## Interesting facts

This species is the only representative of the genus *Naufraga*. While the plant can reproduce from seed, it can also form new individuals vegetatively from its lateral shoots. The flowers have been shown to be pollinated by ants, which is very rare in the plant kingdom. By having tiny flowers grouped close together and near the ground means that ants can quickly visit them to seek nectar, and thus transfer pollen from one flower to another.

## Why is it threatened?

This species has been categorized CR (Critically Endangered) according to IUCN Red List Criteria B1ab(v)+2ab(v). This means that its range is very limited, the population is fragmented, and the number of mature individuals is in decline.

Small, isolated subpopulations cover an area of less than 1,000 m$^2$, which makes this species very vulnerable to extinction. *Naufraga balearica* is sensitive to droughts, mainly in spring. Repeated droughts over the last 20 years have resulted in a continuous decline in the numbers of individuals. With climate change, a scenario of a warmer, drier regime puts this species at risk. Some other species living alongside *Naufraga balearica* are more drought-resistant, thus have benefited from drier conditions and provide increased competition.

*Naufraga balearica* is also threatened by intensive trampling by goats, although grazing may also reduce competition from other species. In the 1980s plants were removed by collectors, which may explain the decline in the original population.

## What is being done to protect it?

**Legally:** This species is listed in Annex I (in danger of extinction) of the Spanish Royal Decree 439/1990, which grants it protection in its natural habitat. Internationally *Naufraga balearica* is included in two legal documents: Appendix I of the Bern Convention and Annexes II and IV of the EC Habitats Directive as a priority species.

*In situ*: In 1997 within the framework of a EU LIFE project entitled "Conservation of natural habitats and plant species in Corsica", several projects were undertaken which included habitat protection, land acquisition, and restoration work for this species. A re-introduction attempt on Corsica using material from the Geneva Botanical Garden was unsuccessful. On the Balearic Islands, a conservation programme undertaken by the Universitat de les Illes Balears and financed by the MAVA Foundation was launched in 2003.

*Ex situ*: Material collected from the Balearic Islands is being cultivated in the Botanical Garden of Sóller on Majorca (Spain). Corsican material (all of the same provenance) has been cultivated in the botanical gardens of Geneva, Brest and Porquerolles since 1981, the year when the population on Corsica was discovered.

## What conservation actions are needed?

Research is needed to understand the reproductive biology and environmental constraints for this species in order to undertake better management. Permanent plots are needed to monitor population numbers over time and the effect of climate change. Research will also help guide re-introduction attempts to Corsica, where ideally Corsican material should be used. In addition, the management and ownership issues of the sites where it either grows or once grew need to be resolved to guarantee the long-term survival of this species.

## Scientific coordination

Dr Maurici Mus, Dpto. Biologia, Universitat de les Illes Balears, Palma de Mallorca, Spain.
Dr Juan Rita Larrucea, Dpto. Biologia, Universitat de les Illes Balears, Palma de Mallorca, Spain.

# *Medicago citrina*

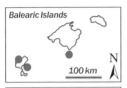

*Balearic Islands*

100 km  N

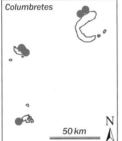

*Columbretes*

50 km  N

HERBARI VIRTUAL DE LES ILLES BALEARS, HTTP://HERBARIVIRTUAL.UIB.ES

MANUEL B. CRESPO

| | |
|---|---|
| Latin name: | *Medicago citrina* (Font Quer) Greuter |
| Common names: | **Alfalfa arbórea (Spanish); Alfals arbori (Catalan); Moon Trefoil, Tree Medic (English)** |
| Family: | **Papillionaceae (legume family)** |
| Status: | **CRITICALLY ENDANGERED (CR)** |

## Where is it found?

This species only grows in small, fragmented populations on the rocky slopes of some Balearic Islands, the Columbretes archipelago (province of Castellón), and on one small islet (the 'Illot de la Mona' or 'Escull del Cap de Sant Antoni') situated just off the coast of the Cape of St Antoni (province of Alicante). There are no known populations on the large Balearic Islands (Ibiza, Cabrera and Majorca) themselves, only on the islets surrounding Ibiza and Cabrera. Plants have been introduced onto one islet of Majorca.

## How to recognise it

*Medicago citrina* is a shrub or small tree reaching 2 m in height, with an erect trunk. It has alternate, compound leaves composed of three rounded leaflets. Flowers are bright yellow in colour, initially forming dense bunches but becoming more spread out as the fruits start to mature. The plant flowers during the spring and the fruits form spiral-shaped pods.

## Interesting facts

The plant grows on several small islands, and is thought to be dispersed by seabirds or other animals. It seems that seed germination improves after passing through an animal's digestive tract.

## Why is it threatened?

This species has been categorized as CR (Critically Endangered) according to IUCN Red List Criteria B1ab(iii,v)+2ab(iii,v). Reasons include a total growing area of less than 10 km², a very fragmented global population, and a decline in number of mature individuals as well as quality of habitat. In 1997, the population on the Columbretes decreased by 40% due to an attack of scale insects (*Icerya purchasi*) native to Australia; the same insects were detected on the Balearic Islands in 2001. Other threats include introduced rabbits, alien species such as *Opuntia maxima*, and periodic, severe attacks of the parasitic plant *Cuscuta*. Invasive species may be more tolerant of drought than *Medicago citrina*, which has a considerably reduced fruit set under dry conditions.

## What is being done to protect it?

**Legally:** The species is presently listed in Annex I of the Spanish Royal Decree 439/1990, as 'sensitive to the disturbance of its habitat', which guarantees protection of its natural habitat. Since 1985 it has been strictly protected within the region of Valencia by a regional government decree.

*In situ*: On the Columbretes, *Medicago citrina* occurs within a federally owned nature reserve. Access to populations on the islands Ferrera and Foradada is strictly prohibited apart from scientific expeditions. These two islands have now been designated as micro-reserves, with a management plan developed in 1993. The species has been re-introduced to the island of Grossa where it was eradicated by rabbits brought to the island in the 18th and 19th centuries. The rabbits were eliminated by 1987.

On 'Illot de la Mona', approximately 25 plants grow within a plant micro-reserve, itself within the boundaries of El Montgó Nature Park. An action plan for the micro-reserve has recently been approved.

On the Balearic Islands most of the populations occur within the National Park of Cabrera. The Botanical Garden of Sóller has a programme for the islets around Cabrera, including re-introduction and monitoring. It is intended that these measures will also be applied to the islets of Ibiza. All the small Balearic Islands are protected as Natural Areas of Special Interest under the Parliament of the Balearic Islands Law 1/1991.

*Ex situ*: This plant is cultivated and seeds stored in the Botanical Garden of Valencia. It is also cultivated in the Botanical Garden of Sóller and IMIDA (Instituto Murciano de Investigación y Desarrollo Agropecuario) of Murcia.

## What conservation actions are needed?

The priority is to control attacks by the scale insect. It has been noted that the scale insect is a problem in citrus farms on the Spanish mainland, since farmers spray their citrus trees to kill a leaf miner, and in doing so kill a ladybird which is the main predator of the scale insect. If correct, measures are needed to maintain ladybirds in the areas where *Medicago* grows. A method of controlling *Cuscuta* also needs to be identified, and invasive alien species (e.g. *Opuntia maxima*) need to be managed. In general, conservation efforts require more information about the population trends of *Medicago citrina* over a prolonged period. Re-introduction work needs to be continued.

## Scientific coordination

Dr Manuel B. Crespo, CIBIO (Instituto Universitario de la Biodiversidad), Universidad de Alicante, Spain.
Dr Ana Juan, CIBIO (Instituto Universitario de la Biodiversidad), Universidad de Alicante, Spain.
Dr Maurici Mus, Dpto. Biologia, Universitat de les Illes Balears, Palma de Mallorca, Spain.

# Anchusa crispa

FEDERICO SELVI

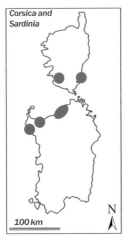

Corsica and
Sardinia

100 km

N

FEDERICO SELVI

| | |
|---|---|
| Latin name: | *Anchusa crispa* Viv. |
| Common name: | Buglosse crépue (French) |
| Family: | Boraginaceae (forget-me-not family) |
| Status: | CRITICALLY ENDANGERED (CR) |

### Where is it found?

This species is only found on the upper part of beaches on the islands of Corsica and Sardinia, where it grows on sandy soils. It is known from 14 sites on the north-western coast of Sardinia and six sites on Corsica, altogether covering an area of less than 10 km².

### How to recognise it

This small annual or short-lived perennial plant varies between 10-35 cm in height, branching from the base. The upright stem later trails on the ground, and is partially covered by long, stiff, almost spiky hairs. Its leaves are 5-10 cm long, lance shaped, and finely toothed. The small flowers are grouped in loose, alternate inflorescences and open between March and June. Flowers in the same inflorescence have different colours depending on their age, ranging from pale blue to blue-violet.

### Interesting facts

*Anchusa crispa* depends on the substrate on which it grows (depth, richness and humidity) in order to accumulate sufficient energy reserves to flower and fruit. Ants disperse its tiny seeds over short distances, and they may be dispersed over longer distances by cows (which spread undigested seeds) and by water. *Anchusa crispa* is usually found growing on fairly firm sandy substrates at the upper edge of the beach, sometimes associated with sand couch grass (*Agropyron junceum*) and marram grass (*Ammophila arenaria*). Although *Anchusa crispa* tolerates occasional trampling, it will disappear if the pressure becomes too severe.

## Why is it threatened?

This species has been categorized CR (Critically Endangered) according to IUCN Red List Criteria B1ab(iv)c(iv)+2ab(iv)c(iv). This means that the area in which it is found is very small, all the populations are fragmented and declining, and the number of mature individuals fluctuates severely.

Even if 20 populations are still known to exist on Sardinia and Corsica, these are all very small. At least four on Corsica (two sites to the north of the Gulf of Valinco and two on the eastern coast) are in steep decline, and another site at Campitellu disappeared in 1999. Up to now conservation measures do not seem to have effectively stopped the decline. Fragmentation of all the populations makes them particularly vulnerable.

The threats to this species can be divided into two groups: those posed by people, and those by natural events. Currently this species is badly affected by human activities, in particular intense trampling; motorbikes; four-wheel drive vehicles and quad bikes; camping; the construction of tracks and roads; mechanical beach cleaning; and the removal of sand. The construction of ditches upstream to the beaches also poses a threat by modifying the amount of water available to the plants.

On Corsica, severe storms have repeatedly swept large amounts of sand over the area where *Anchusa* grows. In 1999 and 2002 such events caused substantial declines in several populations of this species.

## What is being done to protect it?

**Legally:** Legally: In Corsica, *Anchusa crispa* is legally protected at the national level and included in the French Red Book (*Livre Rouge des plantes menacées de France*). In Sardinia it is included in the Italian Red Book (*Libro Rosso delle piante d'Italia*) where it is listed as Vulnerable. Internationally, it is included in Appendix I of the Bern Convention, and as a priority species in Annexes II and IV of the EC Habitats Directive.

*In situ*: Several conservation projects have started through a LIFE project entitled "Conservation of natural habitats and plant species in Corsica", including habitat protection, land acquisition, and restoration work.

Under the EC Habitats Directive, several sites containing *Anchusa crispa* are now included in the European Natura 2000 network, a measure likely to strengthen conservation efforts. Certain sites on Corsica are being restored by l'Antenne Corse du Conservatoire Botanique National Méditerranéen de Porquerolles and the Conservatoire Régional des Sites de Corse, which includes management and replanting.

*Ex situ*: Seeds of this species have been conserved in several seedbanks, e.g. at Porquerolles (France), and plants have been cultivated in several botanical gardens, including Sóller (Majorca), Porquerolles, and Geneva (Switzerland).

## What conservation actions are needed?

Beach managers as well as the public need to learn about the importance of protecting native coastal species. Barriers should be erected to keep vehicles off the beach. Sites hosting important populations of *Anchusa crispa* need to be acquired by nature conservation managers.

## Scientific coordination

Dr Jacques Gamisans, Laboratoire de Botanique et de Biogéographie de l'Université de Toulouse, France.
Dr Federico Selvi, Dipartimento di Biologia Vegetale, Università degli Studi di Firenze, Italy.
Dr Guilhan Paradis, Ajaccio, Corse, France.

# Biscutella rotgesii

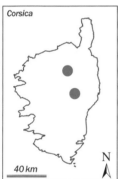

Corsica

40 km

N

| Latin name: | *Biscutella rotgesii* Foucaud |
|---|---|
| Common name: | Lunetière de Rotgès (French) |
| Family: | Cruciferae (mustard family) |
| Status: | CRITICALLY ENDANGERED (CR) |

### Where is it found?

Endemic to Corsica, this species colonizes rocky grasslands and scree on serpentine substrates. Only two populations growing at altitudes of between 150 and 250 m remain: one near Ponte Leccia, and the other in the Inzecca Gorge.

### How to recognise it

This perennial species is 10-25 cm high with an ashy-green stem branched from the base. The edges of its rough, hairy leaves are scalloped and arranged in a basal rosette. Its flowers are pale yellow and gathered in loose bunches, lasting from May to June. The fruits are shaped like eye glasses; when they dry out they open up to release their seeds.

### Interesting facts

Like most species growing on serpentine rock, *Biscutella rotgesii* can tolerate high concentrations of heavy metals in the soil as well as a limited water supply, conditions which limit competition from other species.

### Why is it threatened?

This species has been categorized CR (Critically Endangered) according to IUCN Red List Criteria B1ab(iii,iv)+2ab(iii,iv). This means that the area where it is found is very small and fragmented. Only two populations consisting of at most several hundred individuals exist, and numbers are diminishing.

In addition to such small populations vulnerable to catastrophic effects, road construction, urbanization and fires pose major threats to this species. For example, the fire in 1993 destroyed most of the Ponte Leccia population, which may now have even disappeared.

### What is being done to protect it?

**Legally:** This species is legally protected at the regional level and included in the French Red Book (*Livre Rouge des plantes menacées de France*). Internationally it is not protected by any conventions.

*In situ*: *Biscutella rotgesii* is indirectly protected because most specimens grow within the "Défilé de l'Inzecca", which is a Natura 2000 site. A management plan for this site is being developed by the Direction régionale de l'environnement. A conservation programme for this species is being undertaken by the Parc naturel régional de Corse, supported by the MAVA Foundation.

*Ex situ*: This species has been cultivated at the Botanical Garden of Porquerolles (France) where seeds are also being stored.

### What conservation actions are needed?

Land managers need to be made aware of the importance of saving this species. A management plan for the Natura 2000 site "Défilé de l'Inzecca" needs to be drafted and implemented. It is also necessary to reinforce the population at Ponte Leccia by *ex situ* propagation from seeds collected on-site. This species should also be cultivated and seeds stored at other botanical gardens. Finally, land should be acquired by nature conservation managers in order to protect areas most at threat due to human activity.

### Scientific coordination

Dr Jacques Gamisans, Laboratoire de Botanique et de Biogéographie de l'Université de Toulouse, France.

# Centranthus trinervis

Corsica

40 km

N

| Latin name: | *Centranthus trinervis* (Viv.) Béguinot |
|---|---|
| Common name: | Centranthe à trois nervures (French) |
| Family: | Valerianaceae (valerian family) |
| Status: | CRITICALLY ENDANGERED (CR) |

### Where is it found?

Endemic to Corsica, *Centranthus trinervis* is only known from a single population growing on the granitic boulders of Trinité near Bonifacio in the south-west of the island. It grows in the shade of rock crevices and along cliff terraces.

### How to recognise it

This herbaceous species has an upright, hollow stem with a woody base. It grows in clumps some 20-60 cm in height. Flowering from April to July, its pink, sometimes white tubular flowers are grouped in bunches, and have five petals with a protruding stamen in the centre. The dry fruit is crowned by feathery hairs to help in wind dispersal. The leaves are in opposite pairs, shiny and elongated, and with three nerves, hence the name of this species.

Even though this species is included in the list of threatened plants of Italy, *Centranthus trinervis* only occurs on Corsica (France). A similar species, *Centranthus amazonum* from Sardinia, has been confused with this one. It can be distinguished from

*Centranthus trinervis* by its narrower dull blue-green leaves and longer fruits. Both species grow in different habitats and have a different flowering period.

## Interesting facts

Only the woody base of this plant persists throughout the entire year, as other above-ground parts dry up at the beginning of summer or are broken by autumn storms. The single population grows in a windswept site, but plants are not exposed to sea spray. After a fire, partially burned individuals may regenerate from small suckers at the base. The fruits are dispersed by wind, although regeneration has not been observed outside its present location.

## Why is it threatened?

This species has been categorized CR (Critically Endangered) according to IUCN Red List Criteria B1ac(iv)+2ac(iv); C2b. This is because only one single population of less than 200 individuals remains, and the number of mature individuals fluctuates severely. A combination of natural and human factors could rapidly lead to the extinction of this species in its natural habitat.

Since this species is very light-demanding, rapid overgrowth of the site by lianas such as Sarsaparilla (*Smilax aspera*) may pose a threat. Fires may kill some individuals, but at the same time, the gaps created may provide a positive short-term effect.

Some plants have been destroyed in several places by rock climbers. The development of a tourist complex and a housing estate at the foot of the cliff, combined with the planting of invasive ornamental plants such as Red Valerian *Centranthus ruber* and Pampas Grass *Cortaderia jubata* may pose other threats to this species. Although *Centranthus ruber* has not yet shown signs of being particularly invasive on Corsica, it could hybridize with *Centranthus trinervis*.

## What is being done to protect it?

**Legally:** This species is legally protected at the national level and listed in the French Red Book (*Livre Rouge des plantes menacées de France*). Internationally, it is included in Appendix I of the Bern Convention and in Annexes II and IV of the Habitats Directive.

**In situ:** This species' habitat is included in the European Natura 2000 network, which in theory affords it protection.

**Ex situ:** Seeds of this species are preserved in several seedbanks (e.g. Porquerolles), and cultivated in the botanical gardens of Brest and Porquerolles (France) and Geneva (Switzerland).

## What conservation actions are needed?

Most urgently, the unique site where this species is found needs to be managed to allow the population to increase. It is essential that the habitat be kept open by clearing competing species such as *Smilax aspera*, and by eliminating the cultivation of *Centranthus ruber* and other invasive species grown in the area. Climbing should also be forbidden on the north-east slope of the Trinité mountain. Climbing associations as well as the general public need to be made aware about the conservation status of rare plant species in the region.

Second, more field studies and regular monitoring are needed to understand the effect of vegetation cover and density, fires and storms on this species. A replanting programme from seeds collected *in situ* and plants grown in botanical gardens should be launched.

## Scientific coordination

Dr Alain Fridlender, Université de Provence, Faculté St-Charles, Marseille, France.
Dr Daniel Jeanmonod, Conservatoire et Jardin Botaniques de la Ville de Genève, Geneva, Switzerland.

# *Limonium strictissimum*

Corsica and Sardinia

100 km

N

Latin name:     *Limonium strictissimum* (Salzmann) Arrigoni

Common name:     **Statice à rameaux raides (French)**

Family:     **Plumbaginaceae (sea lavender family)**

Status:     **CRITICALLY ENDANGERED (CR)**

### Where is it found?

Endemic to Corsica and Sardinia, this species grows near the sea on a variety of substrates: on sand of various coarseness, as well as on granitic or limestone boulders. In the northern part of Sardinia this species is only found on the granitic rocks of Punta Rossa on the island of Caprera (in the Maddalena Archipelago), where only a few dozen individuals are known. On Corsica some 1,200 individuals are known from five sites, forming three subpopulations.

34

## How to recognise it

This perennial plant has a woody base with upright stems, reaching up to 30 cm tall (including the inflorescence). Its leaves are linear or spoon-shaped, 1.5-2.5 cm long by 0.2-0.4 cm wide, and overlap at the base of the stem. Although not small, they are usually difficult to see as they are often covered by dead sea grass (*Posidonia oceanica*) carried ashore by the wind. Inflorescences measuring 10-30 cm long develop during summer, mainly in August. They resemble leafless, upright branches, with tiny bluish, tubular flowers.

Individuals growing on sandy to gravely substrates are usually larger (as well as more numerous) than those growing on boulders, probably due to a better water supply.

## Interesting facts

The genus *Limonium* has been split into a large number of species which are often very difficult to distinguish; around 300 species have been described from Mediterranean shores. *Limonium strictissimum* belongs to the group related to *Limonium articulatum*.

This species is capable of producing seeds without the flowers ever being pollinated, which is known as apomixis. This may be one reason why so many *Limonium* species have evolved.

## Why is it threatened?

This species has been categorized CR (Critically Endangered) according to IUCN Red List Criteria B1ab(iii,v)+2ab(iii,v). This means that its area where it occurs is very small, that the four subpopulations are severely fragmented, and that the habitat extent and quality is decreasing as well as the number of individuals.

This species is threatened by natural factors such as drought and landslides, both along the cliffs and along the little strips of beach where it grows. Various human activities also pose a threat: trampling by tourists threaten all the sites where this species is found, as does the construction of more resorts, especially on the beach of Maora.

The best subpopulations on Corsica occur on the beaches of Maora and Piantarella. Neither of these sites are protected or proposed for the European Natura 2000 network.

## What is being done to protect it?

**Legally:** This species is legally protected at the national level. Internationally, it is included in Annexes II and IV of the EC Habitats Directive as a priority species.

*In situ:* The Sardinian subpopulation on the island of Caprera should be protected as it is found within the National Park of the Archipelago of Maddalena.

*Ex situ:* From August 2004, seeds are being stored at the Conservatoire botanique de Porquerolles (France).

## What conservation actions are needed?

The administrators responsible for managing beach areas need to be made aware of the urgency of conservation issues for this species. The land on which this species grows should be acquired or managed by nature conservation authorities.

## Scientific coordination

Dr Guilhan Paradis, Ajaccio, Corse, France.
Professor Pier Virgilio Arrigoni, Dipartimento di Botanica ed Ecologia Vegetale, Università degli Studi di Firenze, Italy.

# Anthemis glaberrima

RED DATA BOOK OF RARE AND THREATENED PLANTS OF GREECE

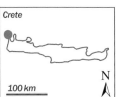

Crete

100 km

N

| Latin name: | *Anthemis glaberrima* (Rech. f.) Greuter |
| Common name: | none |
| Family: | **Compositae (daisy family)** |
| Status: | **CRITICALLY ENDANGERED (CR)** |

### Where is it found?

*Anthemis glaberrima* is endemic to the islets of Agria Gramvousa and Imeri Gramvousa, situated at the extreme northwest tip of the island of Crete, where it grows on littoral rocks. The total area where this plant is found covers less than one square kilometre, with the number of individuals approximately a thousand. Today this plant is only known from Agria Gramvousa. Its continued presence on Imeri Gramvousa needs to be checked.

### How to recognise it

This completely hairless species is an annual, meaning that each year it has to grow from seed stored over the winter. Its stems vary from 2-30 cm tall. The leaves may be entire or deeply incised. The small, daisy-like flower heads are 1 cm in diameter and consist of small tubular yellow flowers in the centre surrounded by white ray flowers (pink underneath). Its dry fruits are ribbed.

### Interesting facts

Rich in endemics, the islet of Agria Gramvousa is one of the few islets that have never been inhabited by domestic or wild ungulates such as goats or sheep. Any introduction of herbivores could threaten the natural balance of this island's vegetation and lead to the extinction of some plant species. *Anthemis glaberrima* belongs to the taxonomic group *Ammanthus* which includes several species (such as *Anthemis ammanthus*, *Anthemis filicaulis* and *Anthemis tomentella*) endemic to the southern islands in the Aegean Sea.

### Why is it threatened?

This species has been categorized CR (Critically Endangered) according to IUCN Red List Criteria B1ab(ii,v)+2ab(ii,v). This means that the area where it is found is very limited, and it is expected that both its extent as well as the number of individuals will decline. The main threats to this species include accidental or deliberate introductions of herbivores or invasive plant species to the islets.

### What is being done to protect it?

**Legally:** Internationally, this species is included in two legal documents: Appendix I of the Bern Convention and Annexes II and IV of the EC Habitats Directive, where it is listed as a priority species. The plant is included in Natura 2000 site GR 4340001, which gives it indirect protection.

*In situ*: No measures taken yet.

*Ex situ*: Some seeds have been deposited in the Mediterranean Agronomic Institute of Chania in Crete.

### What conservation actions are needed?

Given the number of species endemic to Agria Gramvousa, the islet should be designated as a nature reserve, and a management plan put in place to prevent the introduction of herbivores and invasive plants. Seeds of *Anthemis glaberrima* should also be stored in seedbanks and specimens cultivated in botanical gardens.

### Scientific coordination

Professor Gregoris Iatroú, Department of Biology, Division of Plant Biology, Institute of Botany, University of Patras, Greece.

# Bupleurum kakiskalae

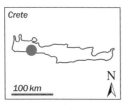

Crete

100 km

N

| Latin name: | *Bupleurum kakiskalae* Greuter |
| --- | --- |
| Common name: | none |
| Family: | Umbelliferae (celery family) |
| Status: | CRITICALLY ENDANGERED (CR) |

### Where is it found?

Endemic to the island of Crete, *Bupleurum kakiskalae* occurs only on a single limestone cliff at Kakiskalo in the mountain range of Levka Ori, above the gorges of Samaria at an altitude of 1,450 m. Approximately 100 individuals are known in the wild.

### How to recognise it

During most of its life, this perennial species produces only leaves, which grow in a basal rosette of about 15-30 lance shaped leaves between 10-25 cm long. After 10-15 years in a vegetative state, it forms a single stem up to 1 m in height. The stem supports one umbrella-shaped inflorescence consisting of small yellow flowers, that open between June and July. After having flowered and produced fruits, the plant dies.

### Interesting facts

Given that only a small number of individuals remain (about 100) and the fact that the plant flowers only once during its lifetime, there are very few individuals which flower at the same time (between 0 and 20, depending on the year). Seed production and the establishment of seedlings is therefore extremely variable. It seems to have very strict ecological requirements as it is only found on a single cliff of a specific type of metamorphous limestone.

## Why is it threatened?

This species has been categorized CR (Critically Endangered) according to IUCN Red List Criteria B1ab(iii,v)c(iv)+2ab(iii,v)c(iv); C2a(ii)b; D. This means that there is only a single small population growing over a very limited area. Both the area where it is found as well as the number of mature individuals are in decline. Additionally, the number of mature individuals present extreme fluctuations since as few as zero to 20 individuals may flower each year. The main threats facing *Bupleurum kakiskalae* are the low probability of genetic exchange within the population due to the small number of individuals flowering at the same time, and cliff instability, as the substrate on which it grows collapses periodically. Goats may also graze any accessible plant.

## What is being done to protect it?

**Legally:** *Bupleurum kakiskalae* is included as a priority species in Annexes II and IV of the EC Habitats Directive and in Appendix I of the Bern Convention. It is also protected by the Greek Presidential Decree 67/81. The plant is included in Natura 2000 site GR 4340008, which gives it indirect protection.

*In situ:* So far, no special conservation measures have been taken at the site. However, this site occurs within the National Park of the Gorges of Samaria, which provides the species some protection.

*Ex situ:* A few specimens are cultivated, and seeds are stored at the Mediterranean Agronomic Institute of Chania on Crete.

## What conservation actions are needed?

Studies on the biology and ecology of this species must be continued to properly define the conservation measures needed. Additional fieldwork is required to determine whether any other populations exist, and to identify cliffs with similar properties to those of Kakiskalo as areas where potential introductions might be undertaken. Regular collection of seeds should be made in order to conserve the widest possible spectrum of genetic material. The site must not only be protected from goats, which have access to certain individuals, but also from botanists who may remove individuals for collections.

## Scientific coordination

Professor Gregoris Iatroú, Department of Biology, Division of Plant Biology, Institute of Botany, Patras University, Greece.
Dr Zacharias Kypriotakis, Technological Education Institute, Heraklion, Crete, Greece.

# Convolvulus argyrothamnos

Crete

100 km

N

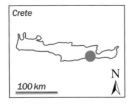

| | |
|---|---|
| Latin name: | *Convolvulus argyrothamnos* Greuter |
| Common name: | none |
| Family: | Convolvulaceae (bindweed family) |
| Status: | CRITICALLY ENDANGERED (CR) |

### Where is it found?

*Convolvulus argyrothamnos* is endemic to Crete and found only on a single limestone cliff in the region of Ierápetra at an altitude of 450 m. Currently fewer than five individuals are known. A second site with approximately 30 plants may recently have been discovered in western Crete, but its status has not yet been confirmed.

### How to recognise it

It is a shrubby plant reaching 80 cm in height, covered by silvery, silky hairs. The linear leaves vary between 1.5-8 cm in length. Its white, bell-shaped flowers, 3.5 cm in diameter, resemble those of common bindweed. They appear in June and are arranged in groups of two to six at the tips of the branches.

### Interesting facts

*Convolvulus argyrothamnos* is one of the numerous species endemic to Crete that currently grow only on cliffs. These species, most of which are evolutionarily ancient, use cliffs as their last refuge. Here they are protected not only from more widespread plants better adapted to current conditions, but also from the voracious appetite of the ever-present goats.

### Why is it threatened?

The species has been categorized CR (Critically Endangered) according to IUCN Red List Criteria B1ab(ii,v)+2ab(ii,v); C2a(i); D. This means that its range is extremely limited as it is only found in one, or at most two, highly fragmented localities. There is a continued decline in number of individuals. Fewer than 50 mature individuals are known, which automatically designates any species as Critically Endangered.

The extremely small number of individuals (eight in 1984, six in 1993 and only four in 1996) at the site near Ierápetra puts this species at a very high risk of extinction. The final blow may be a bush fire, rare-plant collectors, or the great difficulty these plants seem to have reproducing from seed.

### What is being done to protect it?

**Legally:** Internationally, *Convolvulus argyrothamnos* is included in two legal documents: Appendix I of the Bern Convention and as a priority species in Annexes II and IV of the EC Habitats Directive. The plant is included in Natura 2000 site GR 4340002, which gives it indirect protection.

*In situ*: No measures taken yet.

*Ex situ*: Two plants, propagated by cuttings, are currently being cultivated at the Technological Education Institute of Heraklion (Crete). Although they flower, they do not produce fruit. A programme of seed collection and storage at the Mediterranean Agronomic Institute of Chania (Crete) has also been launched.

### What conservation actions are needed?

The survival of this species is unlikely unless it can be rapidly propagated *ex situ* by botanical gardens, followed by reinforcement of the current site(s) *in situ*. It is equally important to protect the site(s) legally, and to undertake effective conservation and management measures.

### Scientific coordination

Professor Gregoris Iatroú, Department of Biology, Division of Plant Biology, Institute of Botany, University of Patras, Greece.
Dr Zacharias Kypriotakis, Technological Education Institute, Heraklion, Crete, Greece.

# Horstrissea dolinicola

Crete

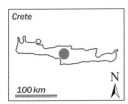

| | |
|---|---|
| **Latin name:** | *Horstrissea dolinicola* Greuter, Gerstb. & Egli |
| **Common name:** | none |
| **Family:** | Umbelliferae (celery family) |
| **Status:** | CRITICALLY ENDANGERED (CR) |

## Where is it found?

This species is only found in the Mt. Ida (Psiloritis) mountain range of central Crete at about 1,500 m above sea level. It grows in a few limestone sinks in a region where many sheep graze during the summer. Its total population numbers just a few dozen individuals in an area of about 3,000 m².

## How to recognise it

*Horstrissea dolinicola* is a perennial species of which more of the plant grows underground than above. Its cylindrical root can be as long as 10 cm, whereas its leaves and inflorescences barely exceed 2-3 cm in height. In spring its deeply incised leaves are the first parts to appear, covering the soil as a rosette. In summer a short stem topped by pinkish flowers, grouped in a round umbrella-like inflorescence, then appears. Fruits are produced in September.

## Interesting facts

This species was first described in 1990. It is the only species belonging to the genus *Horstrissea*, which is closely related to the genus *Scaligeria*. This means that if this species disappears, the entire genus will disappear as well. A great number of species endemic to the mountains of Crete grow in these limestone sinks. All the plants are very small and hug the ground, a strategy which allows them to sustain the grazing pressure of numerous herds of sheep. In the spring, these limestone sinks serve as outlets for melting snow and rain, and are periodically flooded.

## Why is it threatened?

This species has been categorized CR (Critically Endangered) according to IUCN Red List Criteria B1ab(ii,v)+2ab(ii,v); C2a(i); D. This means that the only site where this species is found is very small, and both its range as well as the number of individuals is declining. It is estimated that there may be fewer than 50 mature individuals left in the wild.

*Horstrissea dolinicola* is mainly threatened by over-grazing and nutrient addition by sheep, as well as the possible use of fertilizers. It is also threatened by road construction. At the same time, it seems possible that the sheep may also control other plant species that might compete with *Horstrissea dolinicola*.

## What is being done to protect it?

**Legally:** Currently there is no legal protection for this species. The plant is included in Natura 2000 site GR 4330005, which gives it indirect protection.

*In situ & ex situ*: A replanting programme, supported by the MAVA Foundation, has been undertaken by the University of Patras and the Mediterranean Agronomic Institute of Chania (Crete). This institute also stores seeds of *Horstrissea dolinicola* in seedbanks, but germination has proven to be difficult due to parasites.

## What conservation actions are needed?

It is essential to understand this species' ecology better, especially the role of sheep grazing in order to identify the best steps to take for its conservation *in situ*. It would be reasonable to attempt (re-)introductions of *Horstrissea dolinicola* in other nearby limestone sinks, and to bring this species into cultivation in botanical gardens, as well as store seeds. Finally legal measures for the conservation of this species and its habitat are needed, as well as an awareness campaign for the land-owners and users of this site.

## Scientific coordination

Professor Gregoris Iatroú, Department of Biology, Division of Plant Biology, Institute of Botany, University of Patras, Greece.
Dr Christina Fournaraki, Mediterranean Agronomic Institute of Chania, Crete, Greece.

# *Arabis kennedyae*

C.S.CHRISTODOULOU

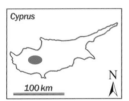

Cyprus

100 km

N

| Latin name: | *Arabis kennedyae* Meikle |
|---|---|
| Common name: | Troodos rockcress (English) |
| Family: | Cruciferae (mustard family) |
| Status: | CRITICALLY ENDANGERED (CR) |

### Where is it found?

This plant grows at 900-1,350m above sea level on the island of Cyprus. It colonizes rocky slopes of the central Troodos and Tripylos mountains (Troodos range). The species is found in semi-shaded, rocky streamside areas dominated by golden oak (*Quercus alnifolia*) and Calabrian pine (*Pinus brutia*).

### How to recognise it

*Arabis kennedyae* is an annual or biennial herb with erect, often purplish stems usually less than 30 cm tall. The basal leaves are up to 6 cm long, 2 cm wide, and form a rosette. The upper leaves are shorter than the basal ones and clasp the stem. The flowers are small and inconspicuous with four white petals, arranged in loose bunches. The plant flowers from April to May.

The fruits are 25-40 mm long and very narrow, almost looking like extensions of the branched stem. At maturity they dry up and split open into two parts that fall off, leaving an inconspicuous central membrane which supports the tiny seeds.

## Interesting facts

It probably depends on the weather conditions whether this plant lives one or two years. In wetter years the plant can survive the summer drought period. The seeds are dispersed either by wind or by floating along water currents. Since this species is annual or sometimes biennial, the number of individuals fluctuates widely from year to year depending on environmental and climatic conditions. These fluctuations make population monitoring difficult.

## Why is it threatened?

This species has been categorized CR (Critically Endangered) according to IUCN Red List Criteria B1ab(iii)c(iv)+2ab(iii)c(iv); C2a(i). This means that the area and quality of its habitat is in decline, the number of plants fluctuates widely, and the remaining tiny populations are very fragmented. Together the three known populations only encompass a few hundred individuals and cover an area of less than 2 km$^2$. One of the populations is subject to human pressures resulting in habitat destruction. A picnic site is located near this single population and military exercises often take place within and around this area.

*Arabis kennedyae* is potentially threatened by road construction or widening. Forest fires are also a threat. Contrary to other annual species in this habitat, its seeds are neither hard-shelled nor heat resistant, so fire could negatively affect the soil seedbank.

## What is being done to protect it?

Legally: This species is protected by the Bern Convention where it is listed in Appendix I. Based on the results of a LIFE Third Countries Project, *Arabis kennedyae* has also been included as a priority species in Annexes II and IV of the EC Habitats Directive. Moreover, it is included in the *Red Data Book for the Threatened Plants of Cyprus*.

*In situ*: The entire population of this species occurs in Troodos National Forest Park and Paphos State Forest, which have been proposed by the LIFE Third Countries project as Sites of Community Importance (SCI) and Special Protection Areas (SPA) in the European Natura 2000 Network. Any future construction projects or military exercises will have to consider the presence of this species. Part of Tripylos Mountain is already a Nature Reserve where species protection laws are enforced.

*Ex situ*: Seeds were collected from cultivated plants at the University of Athens in 1994. They are stored in the seedbank of the Department of Botany at the University of Athens.

## What conservation actions are needed?

Although the entire population occurs within the Troodos National Forest Park, only one subpopulation grows in a Nature Reserve. The areas where the two other subpopulations occur need to be declared as Nature Reserves by the Cyprus Council of Ministers. According to Cyprus Forest Law, flora and fauna is totally protected within a Nature Reserve. Complementing the current legal protection of this species by Forest Law would help because *Arabis kennedyae* would then be prioritized in forest management plans.

## Scientific coordination

Dr Costas Kadis, Research Promotion Foundation, Nicosia, Cyprus.
Mr Charalambos S. Christodoulou, Forestry Department, Ministry of Agriculture, Natural Resources and the Environment, Nicosia, Cyprus.

# *Astragalus macrocarpus*
## *subsp. lefkarensis*

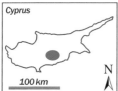
Cyprus

100 km

N

| | |
|---|---|
| **Latin name:** | ***Astragalus macrocarpus* subsp. *lefkarensis*** Agerer-Kirchhoff & Meikle |
| **Common name:** | **Lefkara milk-vetch (English)** |
| **Family:** | **Leguminosae (pea family)** |
| **Status:** | **CRITICALLY ENDANGERED (CR)** |

### Where is it found?

This subspecies only occurs in Cyprus where there are just four populations found near the villages of Lefkara (after which the subspecies has been named), Asgata, Alaminos and Kelokedara. *Astragalus macrocarpus* subsp. *lefkarensis* grows on mountain slopes in the garrigue, which is a low-growing, secondary vegetation derived from the original evergreen mixed forest. It colonizes sunny, dry, calcareous soil.

### How to recognise it

This species is an erect, hairy perennial with a robust stem 30-60 cm high. The leaves are compound with 15-30 leaflets which are longer than wide, and can be opposite or alternate. Its bright yellow flowers are arranged in three to nine clusters, growing from the leaf axils. The flowering period is between March and April. The fruits are egg-shaped and relatively large (2.5-5 cm), containing up to five large seeds.

### Interesting facts

The plant is drought-resistant due to its thick grey-white hairs. The seeds are dispersed when the ripe beans fall off the plant and then roll due to their heavy weight and elliptical shape. No mechanisms for the seeds to be transported uphill are known, but propagation from rhizomes probably prevents the population from gradually being shifted downwards.

### Why is it threatened?

This subspecies has been categorized CR (Critically Endangered) according to IUCN Red List Criteria B2ab(v). This means that its population is severely fragmented and the number of individuals in decline. The typical species *Astragalus macrocarpus*, which occurs in the Eastern Mediterranean and is found in Turkey, is not threatened, while the Cypriot subspecies *lefkarensis* is. Since this subspecies produces a very small quantity of seeds, reproduction is mostly vegetative resulting in very low genetic variability in the four small remaining populations. This may limit its capacity to adapt to environmental changes.

During spring a *Bruchidae* beetle lay its eggs on the flowers. The larvae hatch in the seeds and feed on their nutritive reserves, consuming 50-75% of the seeds. Seed consumption along with fruit and seed abortion are the main reasons for the low reproductive success of this species.

Expansion of tourism around Lefkara brings new threats to the subspecies, including urban development and collectors. A tourist website even notes the occurrence and location of *Astragalus macrocarpus* subsp. *lefkarensis* along a popular hiking trail.

### What is being done to protect it?

**Legally:** This subspecies is protected by the Bern Convention, where it is listed in Appendix I. Based on the results of a LIFE Third Countries Project, it is listed in Annexes II and IV of the EC Habitats Directive as a priority species. It is included in the *Red Data Book for the Threatened Plants of Cyprus.*

*In situ*: In the context of the above EU LIFE project, three sites with populations of this taxon have been nominated as Sites of Community Importance (SCI) by the European Natura 2000 Network.

*Ex situ*: A small number of seeds collected from the Lefkara and Asgata populations are stored in the seedbank of the Department of Botany at the University of Athens.

### What conservation actions are needed?

*In situ*: The reproductive success of this taxon must be increased by reducing seed predators through biological control. Moreover, the Government of Cyprus needs to inform landholders about the presence of this rare taxon on their property and encourage its protection.

*Ex situ*: Since the plants have a very low reproductive capacity, collecting large numbers of seeds from the wild and storing them in a seedbank could affect the recruitment of natural populations. Therefore, any collection for *ex situ* conservation needs to be carried out carefully and documented in order to conserve the maximum genetic diversity in the collections and create the least damage to the wild populations. Cultivation attempts are urgently needed.

### Scientific coordination

Dr Costas Kadis, Research Promotion Foundation, Nicosia, Cyprus.
Mr Charalambos S. Christodoulou, Forestry Department, Ministry of Agriculture, Natural Resources and the Environment, Nicosia, Cyprus.

# *Centaurea akamantis*

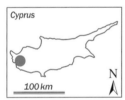

Cyprus

100 km

N

| Latin name: | *Centaurea akamantis* T. Georgiadis & G. Chatzikyriakou |
|---|---|
| Common name: | **Akamas centaury (English)** |
| Family: | **Compositae (daisy family)** |
| Status: | **CRITICALLY ENDANGERED (CR)** |

### Where is it found?

This plant is only found on the Akamas peninsula in the north-western part of Cyprus. It colonizes steep and humid limestone cliffs in the Avakas and Argaki ton Koufon ("Stream of Snakes") Gorges.

### How to recognise it

*Centaurea akamantis* is a semi-woody herbaceous plant with drooping shoots up to 60 cm long. Its leaves are alternate and compound, divided either once or twice. The greyish-green leaflets are linear or spear-shaped, 1-5 mm wide and covered with white matted hair. The small purple-mauve flowers are densely grouped in terminal flower heads that resemble large solitary flowers. The outer flowers of an inflorescence are each reduced to one large "petal"; the inner flowers are tubular and more inconspicuous. The fruits are grain-like with a bunch of fine hairs, which help wind dispersal. It flowers from May to November, producing fruit that ripens between July and December.

## Interesting facts

*Centaurea akamantis* is characterized by an extremely long flowering and fruiting period.

## Why is it threatened?

This species has been categorized CR (Critically Endangered) according to IUCN Red List Criteria B1ab(iii)+2ab(iii). This means that there are only two small populations covering an area of less than 1 km², and the extent and quality of its habitat is declining. One population has only 50 individuals, and the other approximately 500. The populations are isolated from each other and if one of them disappears the other is unlikely to be recolonized.

Increasing visitor numbers to the Akamas peninsula is contributing to a decline in habitat quality, although the number of mature individuals has remained stable since 1993, when the species was first described. Grazing poses a serious threat, even though it is not permitted in these areas and fines are imposed by the Forestry Department.

## What is being done to protect it?

Legally: This species is protected by the Bern Convention where it is listed in Appendix I. It is also listed as a priority species in Annexes II and IV of the EC Habitats Directive. The species is included in the *Red Data Book for the Threatened Plants of Cyprus*. EU accession has had a positive effect on species protection, as management plans are now harmonized with European law.

*In situ*: The Forestry Department is responsible for the site where *Centaurea akamantis* grows and has published a plan aimed at protecting the area. Several other strategies to protect the Akamas peninsula have been proposed, including its designation as a Site of Community Importance (SCI) by the European Natura 2000 Network and the establishment of a National Park. In 2002 a detailed EU/World Bank Action Plan was produced for the proposed park. However, it has not been established due to resistance from local communities. This situation demonstrates the difficulty of reconciling the needs of nature conservation and tourist-based economies, especially in the Mediterranean region.

*Ex situ*: Small numbers of seeds have been collected from the Avakas Gorge and stored in the seedbank at the Department of Botany at the University of Athens. The species has been successfully cultivated at the Cyprus Agricultural Research Institute. However more studies are needed, particularly because it has the potential to be cultivated as an ornamental.

## What conservation actions are needed?

*In situ*: The National Park project should be approved by the Cyprus Council of Ministers. The species' habitat should be nominated as a Nature Reserve, which according to the Forest Law, will provide complete and permanent protection to this site and reduce grazing pressure.

*Ex situ*: Wild seeds of a sufficient genetic range need to be collected and stored in the University of Athens seedbank or other seedbanks. The species would also benefit from cultivation in botanical gardens.

## Scientific coordination

Dr Costas Kadis, Research Promotion Foundation, Nicosia, Cyprus.
Mr Charalambos S. Christodoulou, Forestry Department, Ministry of Agriculture, Natural Resources and the Environment, Nicosia, Cyprus.

# *Delphinium caseyi*

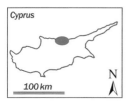

Cyprus

100 km

N

| | |
|---|---|
| **Latin name:** | *Delphinium caseyi* B. L. Burtt |
| **Synonym:** | *Delphinium fissum* **subsp.** *caseyi* C. Blanche & J. Molero |
| **Common name:** | **Casey's larkspur (English)** |
| **Family:** | **Ranunculaceae (buttercup family)** |
| **Status:** | **CRITICALLY ENDANGERED (CR)** |

### Where is it found?

*Delphinium caseyi* is only found in northern Cyprus where it has been recorded at two sites in the northern Pentadaktylos range, one near the St. Hilarion Peak, and the other at the Kyparissovouno Peak. A recent expedition in May 2004 located about 15 individuals on a peak close to St. Hilarion, most of which had been affected by grazing. No other plants were found in the area. This species prefers full sun and grows at the base of rocky cliffs or in the cracks of limestone boulders.

## How to recognise it

*Delphinium caseyi* is an erect, hairy perennial, which can reach 85 cm in height and has a thick rootstock. The basal leaves are radially divided, resembling a palm leaf with a leaf stalk that can be up to 20 cm long. Leaves on the stem are smaller and have shorter stalks. In May or June, a long, thin stem shoots up from the base of the plant, supporting a dozen or more deep violet, long-spurred flowers in dense inflorescences and hairy petals. The flowering period is from June to July.

## Interesting facts

The plants may be propagated by seeds or by subdivisions of the rootstock. All species of *Delphinium* are toxic, like the majority of the buttercup family. The long spur on the flower resembles the beak of a dolphin, the feature that inspired botanists to name the genus *Delphinium.*

## Why is it threatened?

This species has been categorized CR (Critically Endangered) according to IUCN Red List Criteria B1ab(iii,v)+2ab(iii,v). This means that the remaining population is severely fragmented and that there is a continuing decline in the extent and quality of its habitat, as well as a presumed decline in the number of individuals.

This species is probably one of the rarest of Cyprus' endemics. It was estimated that there were less than 500 mature plants left in the two known subpopulations, which together cover less than 2 km$^2$. However a field visit in May 2004 found only 15 individuals, of which 13 had been grazed leaving only two untouched. Whilst grazing is the major threat to this species, its violet flowers make this species very attractive for wild collecting. There is also a potential threat to the subpopulations from nearby military activities and the construction of an antenna in the area.

## What is being done to protect it?

Legally: This species is protected by the Bern Convention where it is listed in Appendix I. Based on the results of a LIFE Third Countries Project, *Delphinium caseyi* has been included as a priority species in Annexes II and IV of the EC Habitats Directive. It is also included in the *Red Data Book for the Threatened Plants of Cyprus.*

*In situ*: The Pentadaktylos mountain range, which encompasses both existing *Delphinium caseyi* subpopulations, has been proposed by the aforementioned project as a Site of Community Importance (SCI) for the European Natura 2000 Network. In addition, both subpopulations grow within the Karmi State Forest, which is protected by the Forest Law from any private interference.

*Ex situ*: No measures taken yet.

## What conservation actions are needed?

*In situ*: It is important to undertake research projects to monitor the population dynamics of this species and assess its biology and ecology. Either a re-introduction or benign introduction project in protected areas is needed.

*Ex situ*: Seeds should be collected and stored in seedbanks. Additionally, the species should be conserved in selected botanical gardens.

## Scientific coordination

Dr Costas Kadis, Research Promotion Foundation, Nicosia, Cyprus.
Mr Charalambos S. Christodoulou, Forestry Department, Ministry of Agriculture, Natural Resources and the Environment, Nicosia, Cyprus.
Dr Yiannis Christofides, Platres, Cyprus.
Mr Christodoulos Makris, Lemesos, Cyprus.

# *Erysimum kykkoticum*

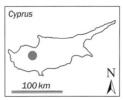

Cyprus

100 km

N

| Latin name: | ***Erysimum kykkoticum*** G. Hadjikyriakou & G. Alziar |
|---|---|
| Common name: | none |
| Family: | Cruciferae (mustard family) |
| Status: | CRITICALLY ENDANGERED (CR) |

## Where is it found?

*Erysimum kykkoticum* is one of the rarest endemics of Cyprus. It is found in the valley of the Xeros River in the western part of the Troodos Mountain range, in a locality called "Argakin tou Pissokremmou" (the stream of Pissokremmos). It grows in fissures of igneous rocks or sometimes on vertical banks of abandoned forest tracks, usually facing east or north, at an altitude of 350-470 m. It is found in association with Calabrian pine (*Pinus brutia*), golden oak (*Quercus alnifolia*) and other shrubs, in habitats characterized by steep slopes and vertical cliffs.

## How to recognise it

*Erysimum kykkoticum* is a glaucous-green woody perennial growing 15-50 cm tall. Its old stems display prominent leaf scars. Its leaves are spoon-shaped, 2-6 cm long, and covered with flattened hairs. Flowering and sterile shoots are also hairy. The inflorescence starts out being densely packed together, later forming a loose bunch of yellow flowers. Its petals are 12.5-14 mm long and slightly hairy on the outside. After flowering an upright, straight or slightly curved, laterally compressed hairy fruit 4.5-8 cm long is produced, containing tiny 4-5 mm oblong seeds. The flowering period begins in mid-March and lasts until mid-May, while the fruiting period lasts from June to July.

## Interesting facts

This species is an evolutionarily ancient member of the *Erysimum* genus, most closely related to the *Erysimum cheiri* group. It colonizes cracks and fissures of rock faces, and as it is a large plant can be confused with some species of *Euphorbia* or *Matthiola* which also grow on rock faces. However in comparison its leaves are relatively large and clearly spoon-shaped.

## Why is it threatened?

This species has been categorized CR (Critically Endangered) according to IUCN Red List Criteria B1ab(v) + 2ab(v). This means that the species grows in a very small area, covering less than three hectares. Based on the results of two inventories carried out in 1998 and 2004, there is a continuing decline of the number of mature individuals and the total population is estimated to number approximately 800 individuals. The major threat to this species' survival is continual forest fires. The species is also potentially threatened by prolonged drought, forestry operations and road construction.

## What is being done to protect it?

Legally: This species grows within the Paphos State Forest which gives it legal protection. It is also listed in the *Red Data Book for the Threatened Plants of Cyprus*.

*In situ*: The entire population of this species occurs in the Paphos State Forest, which has been proposed by the LIFE Third Countries project as a Site of Community Importance (SCI) and Special Protection Areas (SPA) in the European Natura 2000 Network.

*Ex situ*: No measures taken yet.

## What conservation actions are needed?

More research is needed to monitor the population dynamics of this species including its biology and ecology, so that better management plans can be drawn up. In addition, the plant should be brought into cultivation into botanical gardens, and seeds collected and stored in seedbanks.

## Scientific coordination

Dr Costas Kadis, Research Promotion Foundation, Nicosia, Cyprus.
Mr Charalambos S. Christodoulou, Forestry Department, Ministry of Agriculture, Natural Resources and the Environment, Nicosia, Cyprus.
Mr Georgios Hadjikyriakou, Trachoni Lemesou, Cyprus.

# Salvia veneris

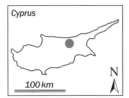

Cyprus

100 km

N

| Latin name: | *Salvia veneris* Hedge |
|---|---|
| Common name: | **Kythrean sage (English)** |
| Family: | **Labiatae (mint family)** |
| Status: | **CRITICALLY ENDANGERED (CR)** |

### Where is it found?

Endemic to northern Cyprus, this species has an extremely local distribution just west of Kythrea. Previously it was believed to be restricted to two small sites at the summit of limestone hills, but more recent fieldwork in spring 2004 found it growing in a continuous area of 12 km², mostly on a particular substrate called Kythrea Flysch (composed of alternating beds of greywacke, marl, sandstone and basal conglomerate). A small part of the population also grows on lava intrusions. It is estimated that there are approximately 4,000 individuals.

## How to recognise it

This perennial plant has a strong woody tap root and felty leaves arranged in a basal rosette. The short flower stems are produced in late March or April. The flowers are bicoloured, the upper lip pale blue, the lower white with pale yellow markings. The shape of the flowers is the only obvious similarity to sage *Salvia officinalis*, the common kitchen herb. The stems produce a faint, pleasant scent when crushed.

## Interesting facts

The arrangement of leaves in a basal rosette is unusual for plants in this family, which normally have leaves in pairs on opposite sides of the stem. The unusual leaf arrangement is thought to be an adaptation to the strong grazing pressure by goats.

## Why is it threatened?

This species has been categorized CR (Critically Endangered) according to IUCN Red List Criteria B1ab(i,iii) which indicates that the area in which the species is found is very small, and that there is a high likelihood of decline in the area, extent and quality of habitat. This species is potentially threatened by any future northward or eastward expansion of the nearby village of Kythrea, which could wipe out the last remaining population. It is also potentially threatened by reafforestation schemes, road construction, military installations and exercises, grazing, burning from the nearby rubbish dump, and dust from nearby limestone quarries.

## What is being done to protect it?

**Legally:** Part of the area in which this species grows lies within the "Lakkovounara State Forest", which is protected by Forest Law from any private interference, while the rest of the area is private or government land. The species is protected by the Bern Convention where it is listed in Appendix I (under the synonym *Salvia crassifolia* Sibth. & Smith). Based on the results of a LIFE Third Countries Project, *Salvia veneris* has been included as a priority species in Annexes II and IV of the EC Habitats Directive. The species is included in the *Red Data Book for the Threatened Plants of Cyprus*.

*In situ*: No measures taken yet.

*Ex situ*: No measures taken yet.

## What conservation actions are needed?

Most importantly, the habitat (particularly that outside the State Forest), as well as the species itself deserves legal protection at the local level. The site should be managed in a way that the species is not endangered by the expansion of the nearby village of Kythrea. Though well-adapted to grazing by goats (and probably not able to compete with faster-growing competitors in the absence of grazing), the site should be managed so that it is not overgrazed. Storage of seeds in seedbanks and *ex situ* cultivation in botanical gardens is recommended. Research is needed to monitor the population dynamics of this species.

## Scientific coordination

Dr Deryck E. Viney, Herbarium of Northern Cyprus, Forestry Department, Alevkaya, Cyprus.
Dr Yiannis Christofides, Platres, Cyprus.
Dr Costas Kadis, Research Promotion Foundation, Nicosia, Cyprus.

# *Scilla morrisii*

C. S. CHRISTODOULOU

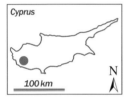

Cyprus

100 km

N

C.S. CHRISTODOULOU

| | |
|---|---|
| **Latin name:** | *Scilla morrisii* Meikle |
| **Common names:** | **Pallid squill, Morris squill (English)** |
| **Family:** | **Hyacinthaceae (hyacinth family)** |
| **Status:** | **CRITICALLY ENDANGERED (CR)** |

### Where is it found?

This plant is found only in the north-western part of Cyprus in three locations. Two of them are near the village of Panagia (Vouni and Aghia Moni monastery) and the third is near the Aghios Neophytos monastery. It grows at an altitude of 250-900 m in moist, shaded crevices and banks, often under a closed canopy of old oak trees (*Quercus infectoria* subsp. *veneris*) and shrubs (*Pistacia terebinthus*).

### How to recognise it

*Scilla morrisii* is a perennial that resembles a small onion. It has three to six thick, linear leaves which emerge from a subterranean bulb and curve around the flowering stalk. The leaves are up to 70 cm long and 0.5-1.5 cm wide. Each of its flowering stems bear one to four flowers, gathered in loose bunches. Flowers are small, star-shaped and lilac or blue, tinged milky-white. The flowering season lasts from March to April.

## Interesting facts

This plant is a perennial which overwinters as a bulb, in which nutrients are stored for the next spring. All species in the *Scilla* genus are known for their toxic properties, which may cause serious digestive disorders.

## Why is it threatened?

This species has been categorized CR (Critically Endangered) according to IUCN Red List Criteria B1ab(i,ii,iii)+2ab(i,ii,iii). This means that its total population is severely fragmented, covering a very small area that is in decline. Today, less than 600 individuals of this species are known, covering an area of less than 2 km². The three known populations are small, isolated from each other, and very sensitive to human pressure.

The survival of this species depends on the conservation of the remaining oak forests. These have been considerably reduced by logging for timber, road construction and expansion of farmland. Large old oak trees have become rare and scattered where they used to form a closed forest cover. While the number of individuals does not seem to be declining, the extent of suitable habitat is decreasing due to road construction and increased agricultural land use.

## What is being done to protect it?

**Legally:** This species is protected by the Bern Convention where it is listed in Appendix I. Based on the results of a LIFE Third Countries Project, the species is also listed in Annexes II and IV of the EC Habitats Directive as a priority species. It is also included in the *Red Data Book for the Threatened Plants of Cyprus*. EU accession has clearly had a positive effect on species protection as management plans are now being rapidly harmonized with European law and strictly enforced.

*In situ*: The two sites where this species is found have been proposed by the LIFE Third Countries Project as Special Protection Areas (SPA) by the European Natura 2000 Network.

*Ex situ*: Small numbers of seeds have been collected from the wild and stored in the seedbank of the Department of Botany at the University of Athens.

## What conservation actions are needed?

*In situ:* The Government of Cyprus should inform landowners of the presence of this rare taxon on their property, and prohibit any action that could threaten these populations.

*Ex situ*: Seeds representative of the genotype of this species need to be collected and stored in several seedbanks. This species should also be grown in botanical gardens.

## Scientific coordination

Dr Costas Kadis, Research Promotion Foundation, Nicosia, Cyprus.
Mr Charalambos S. Christodoulou, Forestry Department, Ministry of Agriculture, Natural Resources and the Environment, Nicosia, Cyprus.

# Aethionema retsina

Greek Islands

200 km    N

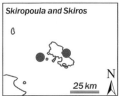

Skiropoula and Skiros

25 km    N

©BOTANISCHER GARTEN UND BOTANISCHES MUSEUM BERLIN-DAHLEM, FU BERLIN

| | |
|---|---|
| Latin name: | *Aethionema retsina* Phitos & Snogerup |
| Common name: | none |
| Family: | Cruciferae (mustard family) |
| Status: | CRITICALLY ENDANGERED (CR) |

## Where is it found?

This species is only known to occur on two Greek islands (the island of Skiros, and the northern part of the much smaller island of Skiropoula, situated just southwest of Skiros). Three populations are known, and are found mainly on the north-east facing cliffs of Mt. Kochilas and some cliffs at Cape Korakia. The species is normally found within half a kilometre of the seashore on vertical limestone rockfaces in small crevices and shady habitats, 10-200 m above sea level.

## How to recognise it

*Aethionema retsina* is a shrubby perennial that forms cushions up to 20 cm high and 40 cm wide. Its woody base can be up to 20 cm thick (i.e. as thick as the length of the shoots). The basal parts produce several shoots per year. These bear smooth, broad, somewhat fleshy leaves 10-15 mm long and rounded at the end. Each young shoot usually develops an inflorescence of 25-50 slightly asymmetric white flowers at its upper end. The petals are up to 3.5 mm long and have a purplish tinge when young. The fruits are flat capsules no more than twice as long as wide. They are framed by a notched wing from which an antenna-like structure (the style) sticks out. Each fruit only contains one lens-shaped seed.

## Interesting facts

This species was not discovered until 1969 and officially described in 1973. It is typical of the specialist plant community adapted to grow in the rocky cracks of seashore cliffs. The plant is fleshy, which is an ecological adaptation to salt spray and common in many seashore plants. Scientists have been undertaking crossbreeding experiments with *Aethionema retsina* and other members of the same family, including plants of enormous agricultural importance such as cabbage, canola and mustard. Scientists are particularly interested in this species' chromosome set and how it relates to that of the cultivated species. One future application could be to genetically improve closely related cultivated species with properties (e.g. salt tolerance and drought resistance) found in *Aethionema retsina*.

## Why is it threatened?

This species has been categorized CR (Critically Endangered) according to IUCN Red List Criteria B1ab(iii,v)+2ab(iii,v). This is because the area in which this species occurs is extremely small and fragmented, and there is severe grazing pressure from increasing numbers of goats, especially during the flowering and fruiting stages.

Due to the small size of the islands, shepherds do not need to fence the territory to keep their herds together, and goats have free access to all parts of the islands. All plants without any efficient grazing protection (spines etc.) are under pressure. *Aethionema retsina* is selectively eaten by goats because of its high nutritional value. A potential threat is limestone quarrying which is very common in the region. This species might easily become extinct at any of the sites where it occurs if stone quarrying starts.

## What is being done to protect it?

**Legally:** This species is not included in any international conventions or national legislation. Due to bureaucratic problems, it has not even been included in the list of Greek species in the European Natura 2000 network.

*In situ*: No measures taken yet.

*Ex situ*: The botanical gardens of Copenhagen (Denmark) and of Lund University (Sweden) have some specimens of this plant in cultivation. However, these do not represent this species' entire gene pool because seeds have only been collected from a few plants.

## What conservation actions are needed?

The priority should be to protect the species from grazing. The areas where it grows should be fenced and managed to keep grazing animals out. Both the large north-east facing cliffs of Mt. Kochilas and the north facing cliffs of the island of Skiropoula should be designated as reserves. Stone quarrying in the species habitat and wild collection of this plant should be prohibited. *Aethionema retsina* is easy to cultivate and has a high potential as an ornamental plant. It should be propagated and planted in other suitable shaded rocky sites, following *IUCN/SSC Guidelines For Re-Introductions* which include guidance on benign introductions.

## Scientific coordination

Professor Gregoris Iatroú, Department of Biology, Division of Plant Biology, Institute of Botany, University of Patras, Greece.

# *Allium calamarophilon*

RED DATA BOOK OF RARE AND THREATENED PLANTS OF GREECE

Greek Islands

200 km

N

Euboea

50 km

N

| | |
|---|---|
| Latin name: | ***Allium calamarophilon*** Phitos & Tzanoudakis |
| Common name: | none |
| Family: | Alliaceae (onion family) |
| Status: | DATA DEFICIENT (DD) |

### Where is it found?

This plant is an endemic to the Greek island of Euboea (or Evia), one of the largest islands in the Aegean Sea. The only known population, comprising just a few individuals, was found in the centre of the island north-east of the small town of Kimi, in 1981. It grows at an altitude of 20-30 m on limestone cliffs that rise almost vertically from the sea.

## How to recognise it

This plant looks similar to an onion. A smooth and unbranched stem 9-13 cm long emerges from the underground bulb. There are one to three leaves wrapped around the lower third or quarter of the stem. They are 1-1.5 mm wide, a similar length to the stem, hairless, and covered by fine canals. The flowers range from white to pink in colour, occurring in bundles of usually five to eight at the top of the stem where they are joined at practically the same point with short stems of equal length. Each flower has five or six "petals" that are joined at the base. One stamen is attached to the interior of each "petal" base and the ovary is slightly heart-shaped. The species flowers in July.

## Interesting facts

Some scientists consider this species to belong to the Lily family (Liliaceae), which includes both onion- and lily-like species. Others prefer to split this large group into more clearly-defined smaller families (for example, the onion family or Alliaceae). Currently the standard taxonomy that is being followed in this book treats the onion family as separate from other lily-like plants, although other treatments lump all these species together into one large family.

## Why is it threatened?

This species has been classified as DD (Data Deficient) according to the IUCN Red List Criteria, meaning that there is inadequate information to make a direct or indirect conservation assessment. Despite being classified as DD, it must be noted that this species is still of very high conservation concern, and could well be Critically Endangered. The species has not been seen since its first description in 1981 and more data is needed to evaluate its status properly. The inaccessibility of its natural habitat has meant that little is known about its true population size and distribution and there is no data available to assess population stability. *Allium calamarophilon* has been included in the "Top 50" to illustrate the difficulties of assessing species which are very poorly known.

One potential threat to this species is a recent plan to build an access road near the seashore. This illustrates the great importance of undertaking environmental impact assessments before building roads or making other modifications to the environment, as road construction could inadvertently destroy the last remaining habitat of this species.

## What is being done to protect it?

**Legally:** This species is not included in any international conventions or national legislation.

*In situ*: There are no current measures in place. In fact, no species research or monitoring has taken place since the original postgraduate project in 1981.

*Ex situ*: *Allium calamarophilon* has been cultivated at the Experimental Botanical Garden of the University of Patras as part of a genetics research project, although due to recent financial constraints the project has stopped and the species is no longer cultivated.

## What conservation actions are needed?

Monitoring is urgently needed to determine if this species still occurs in the wild. If so, its distribution and conservation status should be assessed and a management plan developed.

## Scientific coordination

Professor Gregoris Iatroú, Department of Biology, Division of Plant Biology, Institute of Botany, University of Patras, Greece.

# Consolida samia

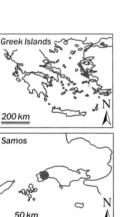

Greek Islands

200 km

Samos

50 km

© ROYAL BOTANIC GARDEN EDINBURGH

| | |
|---|---|
| Latin name: | *Consolida samia* P.H. Davis |
| Common name: | none |
| Family: | Ranunculaceae (buttercup family) |
| Status: | CRITICALLY ENDANGERED (CR) |

### Where is it found?

*Consolida samia* is a Greek endemic that has been reduced to only one remaining population, restricted to the Island of Samos. It grows on the steep limestone scree of Mt. Kerkis at an altitude of 800 m. This species grows in a very specialized habitat: on a gravel-like substrate, with the largest pebbles measuring about 2 cm in diameter. It has never been found in adjacent rocky areas or neighbouring fields with larger stones. The scree on which it grows has no closed vegetation cover, yet despite growing in the open, this plant is easy to overlook because of its very small size.

### How to recognise it

*Consolida samia* is a small annual plant with a 4-6 cm tall, scarcely branched stem covered by dense woolly hairs. Its leaves are divided into three narrow 1 cm long segments. The plant also has smaller hairy modified leaves that are linear and often undivided. It produces one or two characteristic pale lilac flowers per branch. The petals are joined, distinguishable only as three lobes. The middle lobe is short, broad and purple-veined and the two lateral lobes small and triangular. The outer petal has been modified into a 15-17 mm long spur.

## Interesting facts

This species is closely related to *Consolida hellespontica*, but the two species live in distinct habitats and have different ecological requirements. The habitat of *Consolida samia* supports only a few species, most of which are endemic to the Aegean Islands, making it of high scientific interest. The presence of this species indicates that environmental conditions on Mt. Kerkis have remained unchanged for a long period of time.

## Why is it threatened?

This species has been categorized CR (Critically Endangered) according to IUCN Red List Criteria B1ab(iii,v)+2ab(iii,v). In 1975, there were 20 plants at this unique locality and in 1996 their number had reached 100. Now, however, this species cannot be found anywhere. Since it is an annual, it would not be unusual if its population size fluctuated from year to year and its site regularly shifted. The plant may have simply been overlooked due to its small size or just exists in the soil seedbank, awaiting suitable conditions to germinate. The worse case scenario would be that this species is already Extinct.

Re-establishment of this species will become increasingly unlikely over time. Only living specimens are able to disperse their minute seeds uphill by wind. Any potential soil seedbank risks being transported downhill by erosion or stone avalanches, into fields of larger stones which is an unsuitable habitat for this species. Its restricted area of occurrence makes this species sensitive to any habitat modification. If the plant is not extinct, it is potentially threatened by collectors and genetic isolation due to its small population size.

## What is being done to protect it?

**Legally:** This species is listed in Appendix I of the Bern Convention and as a priority species in Annexes II and IV of the EC Habitats Directive. The plant is included in a Natura 2000 site GR 4120003, which gives it indirect protection.

*In situ*: There are no current measures in place.

*Ex situ*: There are no current measures in place.

## What conservation actions are needed?

Field studies are necessary to update current knowledge of the plant's distribution and threats. Studies should include research into its reproduction biology that has not yet been undertaken. Conservation action should be carefully planned based on the results of this research. Expansion of the existing Mt. Kerkis Reserve to include the scree fields would provide further protection and can be easily justified by the presence of other local endemic and endangered plants. The area is also home to rare cliff plant communities, pastures and forests. This species would also benefit from *ex situ* cultivation in botanical gardens.

## Scientific coordination

Professor Gregoris Iatroú, Department of Biology, Division of Plant Biology, Institute of Botany, University of Patras, Greece.

# *Minuartia dirphya*

Greek Islands

200 km

Euboea

50 km

| | |
|---|---|
| **Latin name:** | ***Minuartia dirphya*** Trigas & Iatroú |
| **Common name:** | none |
| **Family:** | Caryophyllaceae (carnation family) |
| **Status:** | CRITICALLY ENDANGERED (CR) |

### Where is it found?

*Minuartia dirphya* grows on the northern slopes of the 1,745 m high Mt. Dirphys in the centre of the Greek island of Euboea (Evia). It is only known from a single population of less than 250 individuals. This species has a narrow geographical range, growing at an altitude of 900-1,000 m. *Minuartia dirphya* grows on serpentine substrates, preferring a thin, infertile soil layer with a high content of rock and gravel and open vegetation coverage (20-40%).

## How to recognise it

This perennial plant forms loose mats or cushions. It is woody at the base and hairless. The non-flowering shoots are up to 4.5 cm long. The tiny rigid leaves are 2-11 mm long, linear with pointed tips and slightly rough margins, growing closely together. The flowering stems differ in appearance from the others in that they bear only 8-14 pairs of leaves, which vary in size, shape and texture. The inflorescences are composed of up to 11 white flowers growing in loose bunches (rarely occurring singularly), supported by small modified leaves that are often tinged purple. The flower stalks measure 1.5-8 mm in length, and the stamens are inserted into a fleshy, light green disc. The flowering period lasts from mid-June to September. The fruit is a capsule of 4-6 mm long and is slightly longer than the sepals.

## Interesting facts

*Minuartia dirphya* was first described in 2000, growing together with *Juniperus oxycedrus* subsp. *oxycedrus* and *Genista acanthoclada*.

## Why is it threatened?

This species has been categorized CR (Critically Endangered) according to IUCN Red List Criteria B1ab(iii,v)+2ab(iii,v). This is because the species is known from a single population covering an area of no larger than 5 km$^2$. Between 100 and 150 individuals have been recorded, although there may be up to 250. The small number of mature individuals, its limited habitat, and the threats from grazing all indicate that the population will decline. The species is severely threatened by grazing from goats and sheep, and from fires made by shepherds.

## What is being done to protect it?

**Legally:** This species is not included in any international conventions or national legislation.

*In situ*: There are no current measures in place.

*Ex situ*: One or two specimens are currently cultivated at the University of Patras. These however do not represent the gene pool of the whole population.

## What conservation actions are needed?

This species needs to be protected from grazing. Its habitat should be fenced and managed to keep grazing animals out. *Minuartia dirphya* should be monitored over a longer period of time to assess population changes and stability. It would benefit from *ex situ* cultivation and its seeds should be collected and stored in seedbanks.

## Scientific coordination

Professor Gregoris Iatroú, Department of Biology, Division of Plant Biology, Institute of Botany, University of Patras, Greece.

# Polygala helenae

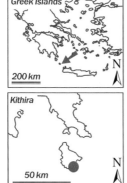

Greek Islands

200 km  N

Kithira

50 km  N

| | |
|---|---|
| Latin name: | *Polygala helenae* Greuter |
| Common name: | none |
| Family: | Polygalaceae (milkwort family) |
| Status: | CRITICALLY ENDANGERED (CR) |

### Where is it found?

This species is endemic to Kithira, a small Greek island of 284 km² which lies opposite the eastern tip of the Peloponnese, Cape Malea, in the Ionian Sea. The taxon is known from a population near Kalamos, but it may also occur in other sites on the island. It is difficult to find not only because it is rare, but also because it is rather inconspicuous. *Polygala helenae* grows on sandy soil and rarely occurs in open areas. Usually its slender branches can be seen protruding from spiny cushions composed of other plant species growing in the same vegetation type.

## How to recognise it

*Polygala helenae* is a low perennial herb. The stem is woody at the base, branched, and covered with fine hairs further up. The lower portion of its stem is flattened along the ground while its upper parts are curved upwards. The lower leaves are longer than broad with nearly parallel sides, becoming longer and narrower higher up. Its inflorescence is composed of 5-30 flowers on short stalks at the top of the stem. The flower stalks are about as long as the modified leaves originating at their base. The flowers have two different types of sepals; the outer two are whitish, 9 mm long and 4 mm wide, and shaped like wings. The petals are pale blue and slightly longer than these wings. The upper petals are distinctly longer than the lower ones. The fruits are capsules containing dark brown seeds about 4 mm long.

## Interesting facts

*Polygala helenae* has a narrow ecological range. It usually grows in association with Spiny Broom (*Genista acanthoclada*) and Thorny Burnet (*Sarcopoterium spinosum*) which typically characterize "phrygana" vegetation. This vegetation type is composed largely of spiny or aromatic dwarf shrubs, growing in lowland areas on dry soils. In Greece there are several types of phrygana, depending upon grazing pressure, the incidence of fires, exposure, soils and geology. A purple tulip, the Greek endemic *Tulipa goulimyi*, may also be found growing in the same areas as *Polygala helenae*. Related species are *Polygala venulosa* and *Polygala supina,* which are Balkan endemics.

## Why is it threatened?

This species has been categorized CR (Critically Endangered) according to IUCN Red List Criteria B1ab(iii)+2ab(iii). This is because it is only presently known from a single population that occurs in a very small area. The natural habitat of *Polygala helenae* was once cultivated. While the area is no longer used for agriculture, there is an increasing risk that it may be needed for agricultural purposes again. Should this happen then the species would probably disappear. Increased tourism also poses a threat to this plant.

## What is being done to protect it?

**Legally:** This species is not included in any international conventions or national legislation.

**In situ**: There are no current measures in place.

**Ex situ**: Cultivation from seeds and attempts to transplant this species from the wild into the Botanical Garden of the University of Patras have both failed.

## What conservation actions are needed?

Given that to date *ex situ* conservation efforts have been unsuccessful, it would seem that the best option to conserve this species is to protect and manage the area where it is known to occur. More fieldwork on Kithira is also needed to see if this species might occur in other areas.

## Scientific coordination

Professor Gregoris Iatroú, Department of Biology, Division of Plant Biology, Institute of Botany, University of Patras, Greece.

# *Saponaria jagelii*

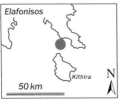

| Latin name: | *Saponaria jagelii* Phitos & Greuter |
|---|---|
| Common name: | none |
| Family: | Caryophyllaceae (carnation family) |
| Status: | CRITICALLY ENDANGERED (CR) |

### Where is it found?

*Saponaria jagelii* grows in the western part of the Greek island of Elafonisos, which is a small island covering about 25 km², located 600 m off the southern coast of the Peloponnese. It is only known to occur on this island where it grows in two scattered, very restricted localities on the sandy sea-shore.

### How to recognise it

*Saponaria jagelii* is an annual plant, which means that it grows each year from seed, rather than resprouting from an existing plant. It has an erect, rather robust branching stem 3-10 cm tall. This stem is reddish, the upper part has some glandular hairs, the lower part almost none. The fleshy leaves are reddish-green, 1-4.5 cm long and lanceolate; the leaf margins are occasionally fringed with fine hairs. The upper leaf surfaces and leaf stalks may be hairy as well. The calyx is cylindrical, reddish, has short teeth, and is covered with glandular hairs. The petals are red with some white at the base, and taper towards the base. The fruit is a nearly cylindrical capsule. The flowering period lasts from March to early June.

### Interesting facts

This species is part of the characteristic plant community growing in disturbed conditions along sandy beaches, together with *Ammophila arenaria*, *Euphorbia paralias*, *Medicago marina*, and *Silene sedoides*. It is related to *Saponaria calabrica*, but differs from the latter in a number of characters.

### Why is it threatened?

This species has been categorized CR (Critically Endangered) according to IUCN Red List Criteria B1ab(i,ii,iii,v)+B2ab(i,ii,iii,v) because the plant is only known to occur at two sites covering a very small area, and the size of the area, quality of habitat, and number of individuals is expected to decline. Tourism is rapidly developing on the island and human activities on the beach, such as driving motor vehicles and trampling, represent a major threat, especially during *Saponaria jagelii*'s flowering period. This increase in tourism could result in a decline of the population or even its extinction within a short time. However, the species should be able to tolerate a moderate number of tourists during the main vacation season after the end of May when its seed capsules are already ripe. At that time, some exposure to trampling could even be beneficial for seed dispersal.

### What is being done to protect it?

**Legally:** This species is not included in any international conventions or national legislation. The plant is included in Natura 2000 site GR 2540002, which gives it indirect protection.

*In situ*: There are no current measures in place.

*Ex situ*: This species is being cultivated in the Botanical Garden of Bochum University, Germany.

### What conservation actions are needed?

Since it is probably not feasible to prevent tourists from accessing the beach, cars at least should be forbidden. Ideally, its habitat should not be accessible during the plant's germination from seeds to the completion of the fruit set, which is from early May to early June. It is also essential to conserve this species *ex situ* by cultivating it in botanical gardens and by storing its seeds in a seedbank.

### Scientific coordination

Professor Gregoris Iatroú, Department of Biology, Division of Plant Biology, Institute of Botany, University of Patras, Greece.

# Cheirolophus crassifolius

ALFRED E. BALDACCHINO

Gozo and Malta

15 km

N

UNIVERSITÀ DI CATANIA

| | |
|---|---|
| Latin name: | *Cheirolophus crassifolius* (Bertoloni) Susanna |
| Synonym: | *Palaeocyanus crassifolius* (Bertoloni) Dostál |
| Common names: | Widnet il-Ba'ar (Maltese); Fiordaliso Crassifoglio (Italian); Maltese rock-centaury, Maltese centaury (English) |
| Family: | Compositae (daisy family) |
| Status: | CRITICALLY ENDANGERED (CR) |

### Where is it found?

Endemic to Malta, *Cheirolophus crassifolius* has a patchy distribution along the north-western and southern cliffs of the islands of Malta, southern Gozo and Fungus Rock. It is confined to coralline limestone seaside cliffs and scree in full sun.

### How to recognise it

This species is a perennial shrub growing up to 50 cm in height, occasionally taller and branching towards the top. Leaves are spatula-shaped and usually smooth and fleshy. Flowering occurs from May to July. Each stalk bears a single flower head, made up of numerous purple tubular florets. The bracts, modified leaves surrounding the flower head, are smooth and without spines or bristles. The plant produces numerous seeds, each equipped with a parachute-like structure to facilitate wind dispersal. *Cheirolophus crassifolius* is distinguished from species of the similar-looking genus *Centaurea* because the latter have spiny bracts and non-fleshy leaves.

### Interesting facts

The Maltese rock-centaury is the National Plant of Malta. This species displays some ancient traits in its habitat preference and flower morphology, and is considered to be a paleoendemic, meaning that it speciated in the distant past and may have been much more widely distributed than today. Previously this species had been placed in a genus of its own (*Palaeocyanus*), but was then grouped with species of the genus *Cheirolophus*. To fully understand the taxonomy of this species, it is important to study its relationship with species of the similar-looking genera *Centaurea* and *Serratula*.

## Why is it threatened?

This species has been categorized CR (Critically Endangered) according to IUCN Red List Criteria B1ab(i,ii,iii,iv,v). This means that the area in which it is found is very restricted (covering less than 100 km$^2$), that the global population is severely fragmented, and that the area where it grows, quality of its habitat, and number of individuals is predicted to decline unless increased conservation measures are taken. The total wild population is estimated at a thousand individuals, but has not been counted.

The species is threatened by a number of factors. First, it is rare to find juvenile plants of this long-lived species, possibly due to the larvae of an unidentified moth observed attacking the developing fruits. Second, the habitat is under threat from quarrying, as fragile boulder cliffs collapse from the pressure wave of nearby dynamite explosions. Dust pollution from quarrying seems to be a minor problem. Third, a number of sites have been affected by human disturbance, especially those most easily accessible. Finally the species, even at inaccessible sites, is threatened by introduced alien plant species, particularly *Opuntia ficus-indica*, *Agave americana* and *Carpobrotus edulis*. These species were originally planted on the plateau but are now invading the cliffs.

## What is being done to protect it?

**Legally:** Internationally, this species is listed in Annexes II and IV of the EC Habitats Directive as a priority species, since Malta's EU accession in May 2004. Nationally, it is protected by the Flora and Fauna Protection Regulations of 1993 and the Flora, Fauna and Natural Habitats Protection Regulations of 2003.

All cliffs on Malta and some cliffs on Gozo are protected locally, either as Sites of Scientific Importance, Areas of Ecological Importance, or Special Areas of Conservation. Fungus Rock (il-Gebla tal-General) is a Strict Nature Reserve. Access is forbidden except for valid scientific reasons.

*In situ*: Management plans are being drafted for a number of sites, including the Qawra-Dwejra Special Area of Conservation (western Gozo).

*Ex situ*: *Cheirolophus crassifolius* has been extensively cultivated, especially since its designation as the Maltese National Plant in 1971. It is now frequently encountered in parks and along the centre-strips of main roads. Note that even if the plant is under cultivation, this does not change its conservation status as the Red List criteria only apply to wild natural populations.

## What conservation actions are needed?

The most effective conservation measures needed are protection and management of the habitat, which means to better control quarrying, prevent illegal dumping (fly-tipping), avoid the introduction of new invasive alien species, and manage the invasives that exist. More cliffs on Gozo need to be protected by law because of their extreme ecological importance. Law enforcement regarding the protection of this species and its habitat needs to be strengthened. Finally, more research is needed to identify the reasons for this species' apparent population decline and habitat fragmentation.

## Scientific coordination

Mr Darrin Stevens, Malta Environment and Planning Authority (MEPA), Sliema, Malta.
Dr Edwin Lanfranco, Department of Biology, University of Malta, Msida, Malta.

# *Cremnophyton lanfrancoi*

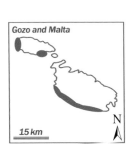

Gozo and Malta

15 km

N

| | |
|---|---|
| **Latin name:** | *Cremnophyton lanfrancoi* Brullo & Pavone |
| **Common names:** | Bjanka ta' l-Irdum (Maltese); Maltese cliff-orache (English) |
| **Family:** | Chenopodiaceae (spinach family) |
| **Status:** | CRITICALLY ENDANGERED (CR) |

## Where is it found?

This species is restricted to the islands of Malta and Gozo. It is rarer than *Cheirolophus crassifolius* (another Top 50 species) and has a similar, but patchier distribution. *Cremnophyton lanfrancoi* grows on sheer seaside cliffs along the north-western and southern cliffs of the islands of Malta and Gozo, including Fungus Rock.

## How to recognise it

*Cremnophyton* is a profusely branched woody shrub between 40-80 cm (rarely 100 cm) in height with dense foliage. The narrow, whitish leaves are somewhat succulent and 2-4 cm long. Flowers are inconspicuous, each subtended by a small modified leaf and arranged in loose terminal bunches. The flowering period lasts from late summer to early autumn. The fruits turn purplish-red as they ripen and develop wing-like structures to facilitate wind dispersal, appear in November.

## Interesting facts

This species has several traits considered ancient in an evolutionary sense, such as an unusual chromosome number (10), and an ecological preference for rock crevice habitats. Like other Maltese endemics, it probably represents a relict element of the old Tertiary flora. *Cremnophyton lanfrancoi* was described in 1987 and is the only species in this genus. It had long been confused with *Halimione portulacoides*, a plant of saline marshlands and dunes (not cliffs) also found on Malta.

## Why is it threatened?

This species has been categorized CR (Critically Endangered) according to IUCN Red List Criteria B1ab(i,ii,iii,iv,v). This means that it grows over a very small area (covering less than 100 km$^2$), that the remaining subpopulations are severely fragmented, and that the area where it grows, quality of its habitat, and number of individuals is predicted to decline unless increased conservation measures are taken. The total wild population is estimated at several thousands, but has not been counted. Some subpopulations have probably disappeared, such as those along the cliffs of the San Pawl il-Bahar-Mistra area on north-eastern Malta.

Very low regeneration has been observed, probably due to an insect (*Eurytoma* sp.) that feeds on the seeds. All wild plants tested were infected by an as yet unidentified fungus that apparently limits reproductive capacity (note that laboratory plants free of the fungus are easy to propagate by cuttings). In its natural habitat, the species is also gradually being replaced by invasive alien plants, particularly *Agave americana*, *Carpobrotus edulis* and *Opuntia ficus-indica*.

Cliff habitats are endangered or have already collapsed due to pressure waves from the explosions of nearby limestone quarrying. Dust pollution from quarrying seems to be a minor problem. A number of subpopulations are directly threatened by dumping of tar and wastes, a crime difficult to control.

## What is being done to protect it?

**Legally:** Internationally, this species is listed in Appendix I of the Bern Convention and since Malta's EU accession in May 2004 in Annexes II and IV of the EC Habitats Directive as a priority species. On the national level, it is protected by the Flora and Fauna Protection Regulations of 1993 and the Flora, Fauna and Natural Habitats Protection Regulations of 2003.

All cliffs on Malta and several on Gozo are locally protected, either as Sites of Scientific Importance, Areas of Ecological Importance or Special Areas of Conservation. Fungus Rock (il-Gebla tal-General), which is located slightly off-shore of the western cliffs of Gozo, is also a Strict Nature Reserve. Access is forbidden, valid scientific reasons excepted.

*In situ*: Management plans are being drafted for a number of sites, including the Qawra-Dwejra Special Area of Conservation (western Gozo).

*Ex situ*: This species has been propagated by laboratory methods with great success and is available to nurseries and gardeners.

## What conservation actions are needed?

It is particularly important to protect more cliffs on Gozo because of their ecological importance. The most efficient conservation action needed is habitat protection and management. Law enforcement should be strengthened, especially, over illegal dumping, collection of wild specimens and the introduction of alien species on the easily accessible plateau. More research is needed to identify the factors responsible for this species' population decline and habitat fragmentation.

## Scientific coordination

Mr Darrin Stevens, Malta Environment and Planning Authority (MEPA), Sliema, Malta.
Dr Edwin Lanfranco, Department of Biology, University of Malta, Msida, Malta.

# Helichrysum melitense

Gozo and Malta

15 km

N

| | |
|---|---|
| **Latin name:** | *Helichrysum melitense* (Pignatti) Brullo, Lanfranco, Pavone & Ronsisvalle |
| **Common names:** | Sempreviva t'G'awdex (Maltese); Perpetuini delle Scogliere Maltese (Italian); Maltese everlasting (English) |
| **Family:** | Compositae (daisy family) |
| **Status:** | CRITICALLY ENDANGERED (CR) |

### Where is it found?

This species is restricted to the western cliffs of the island of Gozo and Fungus Rock Nature Reserve. It is probably extinct on the island of Malta. *Helichrysum melitense* has a patchy distribution and mainly grows on intact limestone coastal cliffs and scree, preferring full sun. Occasionally it may also be found along the more accessible plateau on top of the cliffs.

### How to recognise it

*Helichrysum melitense* is a low shrub with dense foliage, rarely exceeding 50 cm in height. The stems and foliage appear whitish due to a dense cover of woolly hairs. The leaves are spatula-shaped and rounded at the end. This plant usually flowers from May to June producing dense clusters of bright yellow flower heads. Fruiting occurs in summer and early autumn. Its small seeds are attached to a parachute-like structure which helps in wind dispersal. The leaves have an intense aromatic smell.

### Interesting facts

This species is very ornamental and could be cultivated. Several other *Helichrysum* species are used to cure asthma and rheumatism, but there is no evidence of this species being used for medicinal purposes.

### Why is it threatened?

This species has been categorized CR (Critically Endangered) according to IUCN Red List Criteria B1ab(i,ii,iii,iv,v). Only one population of a few thousand individuals remains on Gozo and Fungus rock, covering an area of less than 25 km$^2$. While the number of individuals growing on the inaccessible cliffs seem to be stable, there is a decline on the more accessible plateau.

Cliff habitats are endangered or have already collapsed due to pressure waves from the explosions of nearby limestone quarrying. To a lesser extent, dust from the quarries may also pose a threat. Regeneration of this species is low, possibly due to insects eating the seeds, which jeopardizes the re-establishment of this species where it was once found. Introduced alien plant species pose serious problems, especially *Opuntia ficus-indica*, *Agave americana* and *Carpobrotus edulis*, which are colonizing the cliffs.

Urbanization and tourism are dramatically increasing in this species' habitat on the plateau. Other threats include wild collection for ornamental purposes as well as the recent construction of kiosks and boat-houses near the shore.

### What is being done to protect it?

**Legally:** Internationally this species is listed in Appendix I of the Bern Convention and in Annexes II and IV of the EC Habitats Directive as a priority species, since Malta's EU accession in May 2004. On the national level, it is protected by the Flora and Fauna Protection Regulations of 1993. Part of the cliffs on Gozo is protected locally as a Special Area of Conservation. Fungus Rock (il-Gebla tal-General) is a Strict Nature Reserve. Access is forbidden, valid scientific reasons excepted.

*In situ*: An action plan for this species has been drafted by the Environment Protection Directorate of Malta. The species as well as parts of its habitat are locally protected by Malta's Environment Protection Act and Development Planning Act, which, for example, restricts planting of certain alien species in certain zones. A management plan is being drafted for a key site for this species, namely the Qawra-Dwejra Special Area of Conservation on western Gozo.

*Ex situ*: Some cuttings collected from Dwejra (western Gozo) were planted at the University of Malta's botanical garden for *ex situ* conservation and ornamental purposes. *Helichrysum melitense* has been propagated at the Plant Biotechnology Centre, using micropropagation techniques as part of a joint research programme with the Department of Biology of the University of Malta.

### What conservation actions are needed?

Legal protection of all those parts of the cliffs on Gozo supporting the species is desirable. Legal protection (e.g. against illegal dumping) must be strengthened to protect this species and its habitat. A management plan for the western cliffs on Gozo is needed and it should include the requirements of Natura 2000, to attain sustainable tourism in the area. Efforts should be made to re-introduce the species on the main island of Malta, and *ex situ* propagation for ornamental purposes should be encouraged. More studies are needed to monitor the decline of the specimens on the plateau.

### Scientific coordination

Mr Darrin Stevens, Malta Environment and Planning Authority (MEPA), Sliema, Malta.
Dr Edwin Lanfranco, Department of Biology, University of Malta, Msida, Malta.

# *Aquilegia barbaricina*

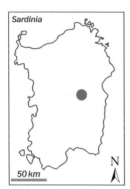

Sardinia

50 km

N

| Latin name: | ***Aquilegia barbaricina*** Arrigoni & Nardi |
| Common name: | **Barbaricina colombine (English)** |
| Family: | **Ranunculaceae (buttercup family)** |
| Status: | **CRITICALLY ENDANGERED (CR)** |

### Where is it found?

*Aquilegia barbaricina* grows in alder scrub along water courses at an altitude of 1300-1400 m and is found only in Sardinia. The 1992 Italian Red List recorded it as growing in a few localities in the central part of Sardinia on Monte Spada, in the Gennargentu area, and in two other localities near Orgosolo. It is now thought to occur only in a few places in a deep wooded valley, which is an exceptional wetland site, on Monte Spada.

## How to recognise it

*Aquilegia barbaricina* is a perennial herb with a single erect, stout stem 30-60 cm high. The stem is covered with fine hairs, branching higher up into three to five nearly leafless flower stalks. The leaf stalks divide once or twice (rarely three times), each division resulting in either a further set of three leaf-bearing stalks, or (at a higher level) three leaflets. Those leaves originating directly from the stem have short stalks and are smaller. Leaflets consist of three lobes with nearly rounded teeth on their outer edge.

The plant has five to eight white, drooping flowers, 25-30 mm in diameter, with five fused petals. Each petal has a slightly curved spur pointing straight upwards. The fruit is an erect capsule, produced in June.

Another "Top 50" species, *Aquilegia nuragica*, resembles *Aquilegia barbaricina* although it has a different coloured flower. In addition, *Aquilegia nuragica* produces fewer flower stalks per plant, its flower spur is more hooked, and its leaflets have sharper teeth than *Aquilegia barbaricina*. It also grows in a completely different habitat.

## Interesting facts

First described in 1977, this species is called a "neoendemic", meaning that it is a species that has evolved relatively recently due to geographic isolation. Its close relative, the "common columbine" *Aquilegia vulgaris* (with pink or purple flowers), is much more widely distributed, but not present in Sardinia.

## Why is it threatened?

This species has been categorized CR (Critically Endangered) according to IUCN Red List Criterion B1ab(ii,v)+2ab(ii,v); D. This means that the area in which it is found is decreasing, and its extremely small population of less than than 50 mature individuals is also in decline.

No long-term study of its population dynamics has been made since the species was discovered relatively recently. However, there is little doubt that the species will become extinct in the near future if no conservation measures are taken. Its rarity and the beauty of its flowers make *Aquilegia barbaricina* attractive for collectors, who can easily access the site. Potential habitat destruction is another threat. Grazing does not seem to be a problem since animals don't like to eat it.

## What is being done to protect it?

**Legally:** Currently there is no legal protection for this species, despite the fact that the regional council of Sardinia proposed a draft law in 2001 concerning protection of plant species on the island, and *Aquilegia barbaricina* is listed in the Annex as an endemic species. However this law is controversial as it may increase collecting interest in the species listed.

*In situ*: There are no current measures in place.

*Ex situ*: No known cultivation attempts have been made.

## What conservation actions are needed?

The areas in which this species occurs naturally should be protected, and all collection should be prohibited. The species also needs to be cultivated in botanical gardens and seeds stored in seedbanks. Better understanding of its reproductive biology and ecology is needed in order to undertake conservation actions including reinforcement or reintroduction.

## Scientific coordination

Professor Ignazio Camarda, Department of Botany and Plant Ecology, University of Sassari, Italy.

# *Aquilegia nuragica*

Sardinia

50 km

N

WATER-COLOUR BY ANNE MAURY IN CAMARDA I. ET AL. (EDS.), 1992, PIANTE DI SARDEGNA

| | |
|---|---|
| Latin name: | *Aquilegia nuragica* Arrigoni & Nardi |
| Common name: | Nuragica columbine (English) |
| Family: | Ranunculaceae (buttercup family) |
| Status: | CRITICALLY ENDANGERED (CR) |

### Where is it found?

*Aquilegia nuragica* is endemic to Sardinia and only found in one area of about 50 m² at Gorropu, near Dorgali. It grows in a gorge along the seasonal Flumineddu river on the nearly vertical lime-stone cliffs. Occasionally it occurs in the dry, sandy pebble substrate of the riverbed as a result of seeds dispersed from the overhanging cliff. However only a few individuals grow here, because they are regularly washed away during floods.

## How to recognise it

*Aquilegia nuragica* is a perennial herb with a single erect, scarcely branching stem 20-45 cm in height, glabrous at the bottom, and covered with fine hairs or glands higher up. All basal leaves have 15-25 cm long, hairless stalks. They have a complex branching system, each division resulting in sets of three (stalks or leaflets). Smaller leaves originating directly from the stem have a less complex branching pattern. Leaflets consist of three segments, not always clearly separated, each with pointed teeth on the outer edge (compared to *Aquilegia barbaricina*, another "Top 50" species, which has leaflets with more rounded teeth).

It has three to five solitary, nodding blue or whitish flowers, 25-30 mm wide at the top. Each of the five fused petals has a hooked spur at the end. The fruit is an erect, drop-shaped capsule with little hooks at the tip.

## Interesting facts

The genus *Aquilegia* has about 70 species worldwide. Three of these (*Aquilegia barbaricina*, *Aquilegia nuragica* and *Aquilegia nugorensis*) are endemic to Sardinia. These and two more (*Aquilegia champagnatii* and *Aquilegia magellensis*) are on the Italian Red List. *Aquilegia nuragica* was first described in 1978. It is clearly genetically isolated from the other Sardinian *Aquilegia* species as well as from the common columbine, *Aquilegia vulgaris*, which occurs all over Europe but not on Sardinia.

## Why is it threatened?

This species has been categorized CR (Critically Endangered) according to IUCN Red List Criterion B1ab(v)+2ab(v); D. This is because the only site where it is known is extremely small, occupying just a few dozen square metres, and the number of individuals is very low and seem to be declining. Only 10-15 individuals are believed to exist in the unique population, although it is difficult to evaluate actual numbers due to the inaccessibility of the site. The lack of other recorded sites makes it difficult to evaluate the status of the decline, but the species is certainly facing a high extinction risk due to natural factors rather than human impact. Grazing does not seem to be a problem since animals do not like to eat it.

## What is being done to protect it?

**Legally:** Currently there is no legal protection for this species, despite the fact that the regional council of Sardinia proposed a draft law in 2001 concerning protection of plant species on the island, and that *Aquilegia nuragica* is listed in the Annex as an endemic species. However this law is controversial as it may increase collecting interest in the species.

*In situ*: The site is situated in the Gennargentu National Park, but due to the lack of a management committee, no protective measures are in place.

*Ex situ*: There are no current measures in place.

## What conservation actions are needed?

The species is mentioned on several tourist websites, which might attract collectors. Detailed information concerning the exact location of the species should be removed in order to keep people away from the site. An action plan urgently needs to be developed and implemented. Cultivation in botanical gardens is recommended.

## Scientific coordination

Professor Ignazio Camarda, Department of Botany and Plant Ecology, University of Sassari, Italy.

# Lamyropsis microcephala

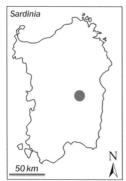

Sardinia

50 km

N

| | |
|---|---|
| **Latin name:** | *Lamyropsis microcephala* (Moris) Dittrich & Greuter |
| **Common name:** | none |
| **Family:** | Compositae (daisy family) |
| **Status:** | CRITICALLY ENDANGERED (CR) |

## Where is it found?

*Lamyropsis microcephala* is endemic to Sardinia, and is found only in the Gennargentu Mountains on the slopes of Mount Bruncu Spina. It grows on eroded rock at 1,500-1,700 m above sea level, in a band of montane dwarf shrub and steppe vegetation.

## How to recognise it

*Lamyropsis microcephala* is a 70-80 cm tall perennial herb resembling a thistle. The plant regenerates every year from buds on a broad rhizome just below the soil surface. The plant has tufts of very spiny whitish stems covered with woolly hairs. The leaves are always longer than their width, sometimes lanceolate in shape, with narrow and deep incised lobed segments that point in all directions. They are covered with stout yellow thorns.

The flower heads are usually solitary, and produced at the top of stems. Bracts (leaves directly beneath the flower head) are spiny and woolly at the tip. The flower head is composed of dozens of tiny tubular white or pinkish flowers, giving the impression of one large flower. Fruits are 5 mm long, smooth, and are especially adapted for wind dispersal, being equipped with a feathery parachute 10-15 mm long.

## Interesting facts

This species has a highly complex branched rhizome system producing many stems, so it is very difficult to distinguish separate individuals. Previously it had been considered possibly Extinct, but was rediscovered shortly before publication of the Italian Red List in 1992.

## Why is it threatened?

This species has been categorized CR (Critically Endangered) according to IUCN Red List Criteria B1ab(iii)+2ab(iii). This is because the species only occurs in a very small area, and the quality of its habitat is declining. The total number of individuals is difficult to determine because of their clonal growth mode. The whole population probably consists of eight to ten colonies, and only covers some 100 m$^2$.

Just to the east of the small area where this species occurs a ski run has been built, which is a serious threat. This species is also threatened by rooting of wild pigs, soil erosion, landslides and further development of the tourist infrastructure. Moreover, this plant has a limited reproductive capacity. The few seeds produced have a very low germination rate, and vegetative growth is very slow.

## What is being done to protect it?

**Legally:** The species is listed in Appendix I of the Bern Convention and as a priority species in Annexes II and IV of the EC Habitats Directive. Nationally there is no legal protection for this species, despite the fact that the regional council of Sardinia proposed a draft law in 2001 concerning protection of plant species on the island, and *Lamyropsis microcephala* is listed in the Annex as an endemic species.

*In situ*: There are no current measures in place.

*Ex situ*: The species is not under cultivation.

## What conservation actions are needed?

A management plan to conserve *Lamyropsis microcephala* is urgently needed, with habitat restoration a priority. The little-used ski run should be removed, and the local population sensitized to the plight of this rare species.

## Scientific coordination

Professor Ignazio Camarda, Department of Botany and Plant Ecology, University of Sassari, Italy.

# Polygala sinisica

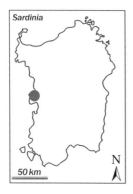

Sardinia

50 km

N

| | |
|---|---|
| **Latin name:** | *Polygala sinisica* Arrigoni |
| **Common name:** | none |
| **Family:** | Polygalaceae (milkwort family) |
| **Status:** | CRITICALLY ENDANGERED (CR) |

## Where is it found?

*Polygala sinisica* occurs only in Sardinia along the coast at Capo Mannu, situated just north of Tharros in the Sinis region. Its entire population is widespread over just a few hectares. It grows in a hot and dry region on a stony or sandy limestone hillside, in vegetation dominated by spiny or aromatic evergreen dwarf shrubs.

## How to recognise it

*Polygala sinisica* is a perennial herb with stems 15-20 cm in length, erect or nearly erect, with some parts trailing along the ground without rooting. Its growth habit is similar to the kitchen herb thyme. The tough, flexible stems are either hairless or slightly hairy. Small, lanceolate or linear, non-fleshy leaves alternate along the stem. 15-20 small pink or bluish flowers on short stalks are grouped in bunches at the ends of the stems. Three of the outer-most petal-like "leaves" of the flower (the sepals) are small and hairy. The two others resemble elliptical wings. The true petals are small, measuring about 11-12 mm long. The plant flowers in May, and immediately afterwards produces fruit with hairy seeds.

### Interesting facts

This species was first described in 1983. It grows in the same community with other Sardinian endemics such as *Arum pictum*, *Genista corsica*, *Ornithogalum biflorum*, *Bellium bellidioides* and *Romulea requienii*. It is found with species of great phytogeographic interest which are also very rare throughout Italy, such as *Helianthemum caput-felis*, *Viola arborescens*, and *Coris monspeliensis*.

### Why is it threatened?

This species has been categorized CR (Critically Endangered) according to IUCN Red List Criteria B1ab(ii)+2ab(ii). This is because the species is only found in a few hectares and it is continuing to decline in the area where it is found. While the exact number of individuals is not known, the native habitat in which this plant grows has been reduced due to current agricultural practices and changes in land use. The land all the way down to the coastline is now farmed, and the hillsides where this species is found are sometimes over-grazed. Other threats include the development of coastal and access roads to the beach and the construction of secondary homes.

### What is being done to protect it?

**Legally:** Currently there is no legal protection for this species, despite the fact that the regional council of Sardinia proposed a draft law in 2001 concerning protection of plant species on the island, and *Polygala sinisica* is listed in the Annex as an endemic species. However this law is controversial as it may increase collecting interest in the species.

*In situ*: Although the Italian Red List of 1992 proposed to set up a nature reserve at Capo Mannu and along the coast of the Sinis peninsula, the site is still threatened by land clearing and receives no legal protection.

*Ex situ*: There are no current measures in place.

### What conservation actions are needed?

The conservation measures suggested in the Italian Red List of 1992 need to be implemented. Full habitat protection, *ex situ* cultivation, and fieldwork to determine the exact size of the population are needed.

### Scientific coordination

Professor Ignazio Camarda, Department of Botany and Plant Ecology, University of Sassari, Italy.

# Ribes sardoum

Sardinia

50 km

N

| | |
|---|---|
| Latin name: | *Ribes sardoum* Martelli |
| Common name: | Sardinian currant (English) |
| Family: | Grossulariaceae (currant family) |
| Status: | CRITICALLY ENDANGERED (CR) |

## Where is it found?

*Ribes sardoum* has been recorded from just one site in Sardinia, occurring in a small south-east facing valley at about 900 m above sea level. The total number of individuals is about 100. The species grows on limestone substrates.

## How to recognise it

This small woody shrub grows up to 0.8-1.5 m tall. It is deciduous, with leaves 1-2 cm wide, ovate or nearly circular, with three to five small lobes and a hairy, glandular leaf stalk. Its small and inconspicuous greenish-yellow flowers, produced around May, are solitary with short stalks. Its fruits are red, ovate berries which ripen between May and June.

## Interesting facts

This species was probably originally distributed throughout the boreal zone, and moved to warmer areas as the climate changed. It is now considered a relict species as it is no longer found anywhere else. Its closest relatives occur in China, Japan, and North America.

Some scientists consider this species part of a larger family, the Saxifragaceae, in which saxifrage- and currant-like species are grouped together. Others prefer to split such large groups up, preferring to work with more clearly defined smaller families (e.g. the currant family Grossulariaceae). In this booklet we have followed the taxonomic standard used at the Royal Botanic Gardens Kew and have attached this species to the currant family.

## Why is it threatened?

This species has been categorized CR (Critically Endangered) according to IUCN Red List Criteria B1ab(v)+2ab(v). This is because the population only covers a small area of around 200 m$^2$, and the number of mature individuals is declining. This decline is due to low seed viability and grazing by goats and sheep. However, despite the grazing pressure, the extent of occurrence seems not to have significantly decreased since *Ribes sardoum* was first recorded at the end of the eighteenth century. However, the lack of historical reference data makes it difficult to evaluate the present population dynamics. Nevertheless the species is seriously endangered by extinction, mainly due to the low seed vitality for reasons not yet clearly understood.

## What is being done to protect it?

**Legally:** The species is listed in Appendix I of the Bern Convention and as a priority species in Annexes II and IV of the EC Habitats Directive. Nationally there is no legal protection for this species, despite the fact that the regional council of Sardinia proposed a draft law in 2001 concerning protection of plant species on the island, and *Ribes sardoum* is listed in the Annex as an endemic species. However this law is controversial as it may increase collecting interest in the species listed. The United Kingdom has included this species in its national legislation [Statutory Instrument 1996 No. 2677. The Endangered Species (Import and Export) Act 1976 (Amendment) Order 1996].

*In situ*: The local population is now committed to protect the site from fire and grazing.

*Ex situ*: *Ribes sardoum* has been cultivated in the Botanical Garden of Florence.

## What conservation actions are needed?

The survival of this species in the wild requires developing serious management plans to protect the species from over-grazing, fire, and plant collection. The species should also be cultivated *ex situ* with the goal of eventually reinforcing the existing population, and perhaps undertaking introductions into similar habitats.

## Scientific coordination

Professor Ignazio Camarda, Department of Botany and Plant Ecology, University of Sassari, Italy.

# *Abies nebrodensis*

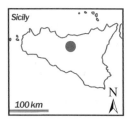

Sicily

100 km

N

| | |
|---|---|
| Latin name: | *Abies nebrodensis* (Lojac.) Mattei |
| Common names: | Abete dei Nebrodi (Italian); Sicilian fir (English) |
| Family: | Pinaceae (pine family) |
| Status: | CRITICALLY ENDANGERED (CR) |

### Where is it found?

Despite its scientific name, the Sicilian fir (*Abies nebrodensis*) does not grow in the Nebrodi Mountains, but is presently limited to the steep, dry slopes of Mt. Scalone in the Madonie Mountains in the north-central part of Sicily. The species is endemic to Sicily and grows at around 1500 m above sea level on limestone soil.

### How to recognise it

*Abies nebrodensis* has a single erect stem, which can reach 15 m high and 60 cm in diameter. In mature trees the crown is broad and conical. The bark is smooth and light grey in young plants, turning orange and becoming thicker, more rugged, and scaly with age.

The needles are short, have flattened cross-sections and variably shaped tips. Their upper surface is dark green and their underside striped blue-green. The needles are arranged in two rows in a horizontal plane and persist during the entire year.

The cones are 8-10 cm long when ripe. They are composed of seed scales, slightly hairy towards their ends and spirally arranged around an axis. Each scale is subtended by a bract with a central short tip. Two light brown winged seeds are inserted at the base of each scale.

This species is closely related to *Abies alba*, but has more resinous buds and shorter needles.

### Interesting facts

The Madonie Mountains, rising to 1,979 m, were once covered by *Abies nebrodensis*. By 1900, the species was considered extinct, due to extensive logging and erosion, but was rediscovered in 1957. Examples of its wood can be seen in the doors and roof-beams of local churches.

### Why is it threatened?

This species has been categorized CR (Critically Endangered) according to IUCN Red List Criterion D (i.e. any plant population numbering fewer than 50 individuals). The current population is around 30 trees, about 23 of which are mature (have produced cones) covering an area of less than 1.5 km$^2$.

Degraded natural habitat, the poor health of specimens propagated in tree nurseries, the limited population size, and threat from fire represent the biggest threats to the species. Hybridization with non-native firs resulting in genetic contamination and global warming also threaten the species.

### What is being done to protect it?

**Legally:** This species is included in Appendix I of the Bern Convention and as a priority species in Annexes II and IV of the EC Habitats Directive. The woodland vegetation in which this species is found is listed in Annex I of the same Directive.

*In situ*: Once rediscovered, foresters immediately initiated conservation measures. However, soil degradation of the natural habitat has made re-introduction difficult. Researchers from Palermo University are currently investigating the species' ideal growth conditions. The species has grown well in several European botanical gardens. An EU LIFE-financed project is being carried out to conserve the existing population. The project includes implementing an action plan which would include forest management, conservation, and the gradual elimination of non-indigenous fir species. The goal is to stabilize the current population and improve the survival rate based on natural reproduction. Its location within the Madonie Regional Park guarantees some level of protection.

*Ex situ*: In 1978, following seed collection, the Forestry Service cultivated 110,000 young trees in a nursery. Since the survival rate in nature is so low, an adoption program was set up in parallel. 40,000 young plants have been planted in the Botanical Garden of Palermo as well as in summer villas and second homes in the Madonie Mountains, a little outside their natural area of distribution. Several mature trees also grow in botanical gardens and arboreta elsewhere in Europe.

### What conservation actions are needed?

For *ex situ* cultivation of *Abies nebrodensis*, areas should be selected that are not home to other fir species to prevent genetic contamination.

### Scientific coordination

Dr Aljos Farjon, Royal Botanic Gardens Kew, Surrey, U.K.
Dr Salvatore Pasta, freelance botanist, Palermo, Italy.
Dr Angelo Troìa, Regional Nature Reserve "Saline di Trapani e Paceco", WWF-Italia, Trapani, Italy.

# *Bupleurum dianthifolium*

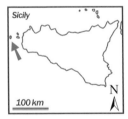

Sicily

100 km

N

| | |
|---|---|
| Latin name: | *Bupleurum dianthifolium* Guss. |
| Common name: | Bupleuro di Marettimo (Italian) |
| Family: | Umbelliferae (celery family) |
| Status: | CRITICALLY ENDANGERED (CR) |

## Where is it found?

This small shrub is endemic to the island of Marettimo (part of the Egadi archipelago, just west of Sicily). It grows in only a few locations on the northern side of the island. It is estimated that approximately 300-500 individuals remain, covering an area of approximately 5 km$^2$. The species grows on calcareous cliffs at an altitude of 20-600 m, preferring north-facing slopes and growing in the cracks of limestone rock faces.

## How to recognise it

This small, cushion-shaped perennial shrub bears its regenerating buds just above the surface of the soil and retains its leaves during winter. Its leaves are crowded at the tip of the branches with herbaceous, almost leafless flowering stems up to 40 cm long. The leaves are linear-lanceolate and often sickle-shaped, being broadest in the middle and hooded at the tip with three to five veins. The compound inflorescence consists of small umbels with three to eight rays. The species usually flowers between May and June.

## Interesting facts

This species is considered to be a paleoendemic, which means that it was once much more widely distributed than today, and probably grew throughout the mountains of the Mediterranean when the region had a tropical climate. The plant reproduces from seeds only, a common characteristic of plants growing in such habitats. Mist is probably its main source of water.

## Why is it threatened?

This species has been categorized CR (Critically Endangered) according to IUCN Red List Criteria B1ab(iii)+2ab(iii). This is because it is restricted to a scattered population on a small islet of 12 km$^2$. Despite growing on largely inaccessible cliffs, it is threatened by grazing wild animals and fire. Given the tiny population and extent of habitat, just one catastrophic event could decimate the remaining population.

## What is being done to protect it?

**Legally:** This species is listed in Appendix I of the Bern Convention.

*In situ*: The whole area of Marettimo Island is included within the SCI (Site of Community Importance) of "Isola di Marettimo" (ITA010002) and within the proposed Marettimo Nature Reserve.

*Ex situ*: Seeds are conserved in the germplasm bank of the Department of Botanical Science of Palermo University. Plants are cultivated in the botanical gardens of Florence and Palermo.

## What conservation actions are needed?

The species would benefit from the entire island being designated as a nature reserve, which would help grazing and fire management. Its collection should be prohibited.

## Scientific coordination

Professor Lorenzo Gianguzzi, Dipartimento Scienze Botaniche, Università degli Studi di Palermo, Italy.
Dr Antonino La Mantia, Dipartimento Scienze Botaniche, Università degli Studi di Palermo, Italy.

# *Bupleurum elatum*

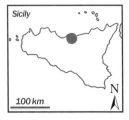

Sicily

100 km

N

**Latin name:** *Bupleurum elatum* Guss.

**Common name:** Bupleuro delle Madonie (Italian)

**Family:** Umbelliferae (celery family)

**Status:** CRITICALLY ENDANGERED (CR)

## Where is it found?

This small shrub is endemic to the Madonie Mountains in the north-central part of Sicily. According to the literature it is found in at least two sites, although only one has been confirmed, with an estimated population of about 400-600 individuals. The inaccessibility of its habitat makes it difficult to estimate the species true distribution, but it is thought to cover about half a square kilometre.

The species grows on calcareous inland cliffs in deep valleys at an altitude between 700 m and 900 m. It prefers north-facing, shady slopes exposed to cold, wet winds, growing on small amounts of soil in rocky cracks.

## How to recognise it

This small perennial shrub has a stout, woody stem of up to 150 cm in length. It bears its regenerating buds just above the surface of the soil, and new stems develop amongst the remains of dead leaves. The leaves vary in shape from twice as long as wide to broadly lanceolate and have 7-9 veins. The basal leaves are arrow-shaped at the base, clasping the stem. The inflorescence is an umbel composed of 6-14 rays. The plant usually flowers between July and August.

## Interesting facts

*Bupleurum elatum* is considered a relict species that once had a much wider distribution. It reproduces only from seeds. Flowering and fruiting have rarely been observed. As is typical of plants growing in such habitats, exposure to mists is likely to be the main source of water.

## Why is it threatened?

This species has been categorized CR (Critically Endangered) according to IUCN Red List Criteria B1ab(iii,v)+2ab(iii,v). This is because it only grows in a very small area and seed production is very low, compromising population regeneration. Although the threats to this species are not sufficiently understood due to its inaccessible habitat, wildfire may represent a serious threat. Given the small and restricted population size, one fire could destroy the remaining population.

## What is being done to protect it?

**Legally:** Although this species is included in regional, national and international Red Lists, it is not protected by any law or convention.

*In situ*: The area of occurrence is included entirely in zone A of the Madonie Regional Park. Note that in the park there are four levels of protection (A, B, C and D), with zone A being the most protected. This zone is very rich in a number of species of high phytogeographic interest.

*Ex situ*: Seeds are conserved in the germplasm bank of the Botanical Science Department of Palermo University, but no known attempts at cultivation have been made.

## What conservation actions are needed?

More research and fieldwork is needed to better understand the distribution and reproductive capacity of this species. Cultivation in botanical gardens will allow researchers to better understand the reasons for its decline in the wild. Collection should be prohibited.

## Scientific coordination

Professor Lorenzo Gianguzzi, Dipartimento Scienze Botaniche, Università degli Studi di Palermo, Italy.
Dr Antonino La Mantia, Dipartimento Scienze Botaniche, Università degli Studi di Palermo, Italy.

# *Calendula maritima*

| | |
|---|---|
| Latin name: | *Calendula maritima* Guss. |
| Common names: | Fiorrancio marittimo, Calendula marittima (Italian); Sea marigold (English) |
| Family: | Compositae (daisy family) |
| Status: | CRITICALLY ENDANGERED (CR) |

### Where is it found?

Endemic to Sicily and some surrounding islets, on the mainland *Calendula maritima* is only found in the Trapani region in a few coastal sites between Marsala and Mt. Cofano. Elsewhere small populations also occur on two or three islets near the Sicilian coast: Isola Grande dello Stagnone, La Maraòne and Favignana (although a recent survey noted that it seems to have disappeared from this site). The species colonizes open nitrogen-rich areas near the sea, and typically grows on decaying remnants of sea grass (*Posidonia oceanica*) washed ashore.

### How to recognise it

This herbaceous plant is 20-40 cm tall and can be woody at the base. Its stems and leaves are covered with short, sticky hairs. The young stems are upright, later branching and drooping to the ground. Unlike the common garden *Calendula*, this species has fleshy leaves with a very strong odour. The shape of the leaves varies from egg-shaped to linear, depending on their position on the stem. This species resembles a little daisy with yellow flowers 3-5 cm in diameter. The main flowering period is May to June, but some flowers can be found throughout the year.

## Interesting facts

This plant is a perennial, but most individuals (possibly due to drought stress) have an annual life cycle. Like annual plants from colder climatic zones, these individuals die back after flowering and setting seed. The seeds then germinate around October. The hot and dry summer marks the end of this species flowering period rather than a cold winter. Other species from the genus *Calendula* are often grown as garden plants. This species has the potential to be developed for horticulture if the optimal culture conditions can be found.

## Why is it threatened?

This species has been categorized CR (Critically Endangered) according to IUCN Red List Criteria B1ab(iii)+2ab(iii). This means that this plant grows in an area of less than 10 km$^2$, that the population is severely fragmented, and its remaining habitat is being reduced. On the main island of Sicily, its natural habitat is under increasing pressure from urban development. The population growing in the well-managed nature reserve "Saline di Trapani e Paceco" is seriously threatened by plans to expand the nearby harbour. The loss of this population would not only diminish the species' gene pool; it would also represent a great loss to science as this locality is where this species was first described (its "*locus classicus*").

In addition, the species is very attractive and may be collected for its beautiful flowers. It is also threatened by competition with an alien invasive species, the "iceplant" or "hottentot fig" (*Carpobrotus edulis*), which grows in part of *Calendula maritima*'s habitat and competes aggressively with it.

## What is being done to protect it?

**Legally:** No measures have been taken to protect the species itself.

*In situ*: Part of the area where this species occurs is situated in the Nature Reserves "Saline di Trapani e Paceco" and "Isole dello Stagnone di Marsala". Here, it is forbidden to collect seeds or any vegetative parts of the plant. These areas are effectively managed (by WWF and the Province of Trapani, respectively), guarded by rangers, and have been subject to scientific monitoring. Construction of roads or houses inside the reserves requires permission. These species-rich reserves are sustainably managed, and economic activities such as salt extraction take place within them.

*Ex situ*: This species is included in the GENMEDOC project (an inter-regional network of Mediterranean seedbanks), and seeds are being collected in order to propagate this species. It should not be difficult to germinate seeds in cultivation, but may be more difficult to meet this species' peculiar habitat requirements (nitrogen-rich, sandy and salty soil).

## What conservation actions are needed?

This species should be added to Appendix I of the Bern Convention and Annexes II and IV of the EC Habitats Directive to give it protection by international law. A campaign to eradicate the invasive species *Carpobrotus edulis* in the region needs to be initiated. *Calendula maritima*'s classical site needs to be protected by finding alternatives to the planned harbour expansion, limiting access to the site, and careful planning of any construction of new roads and buildings. Inventories need to be made over several years to monitor population trends.

## Scientific coordination

Dr Angelo Troìa, Regional Nature Reserve "Saline di Trapani e Paceco", WWF-Italia, Trapani, Italy.
Dr Salvatore Pasta, freelance botanist, Palermo, Italy.

# *Hieracium lucidum*

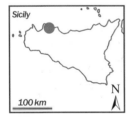

Sicily

100 km

N

| | |
|---|---|
| Latin name: | *Hieracium lucidum* Guss. |
| Common name: | Sparviere di Monte Gallo (Italian) |
| Family: | Compositae (daisy family) |
| Status: | CRITICALLY ENDANGERED (CR) |

## Where is it found?

This plant grows exclusively on Mt. Gallo near Palermo, in north-western Sicily. It is found in only three locations representing just one population in a tiny area. It is not easy to estimate the number of individuals because of the inaccessibility of its habitat. It grows on northern slopes of calcareous maritime cliffs exposed to humid winds. These sites are rich in nitrates from sea bird droppings.

## How to recognise it

*Hieracium lucidum* is a perennial plant with basal leaves that form a radiating cluster on the ground. After flowering the stems die back, but the basal leaves persist. Stems are 10-30 cm in length, with short, star-shaped glandular hairs, particularly near the top. The green leaves have a leathery texture; they are smooth or have a few glandular or simple hairs on the margin. Individual yellow flowers resembling petals are grouped together and look like a single flower, but in reality this is a many-flowered inflorescence. The flowering stem usually has 3-10 such inflorescences, although it can have as many as 40, forming a narrow and compact cluster. It usually flowers between October and November.

## Interesting facts

*Hieracium lucidum* only reproduces from seeds. It is a typical example of a plant which grows in rock crevices and on rock faces. It belongs to a genus containing many very similar-looking species, which presents identification difficulties to the non-botanist (and even many botanists)! The species is very similar to *Hieracium cophanense*.

## Why is it threatened?

This species has been categorized CR (Critically Endangered) according to IUCN Red List Criteria B1ab(i,ii,iii,v)+B2ab(i,ii,iii,v). This is because of the extremely limited area in which this species is found, which is estimated to be 0.2 km². The actual population may only cover as little as 800-1,200 m². Calculations based on these figures estimate that there might only be about 400-500 individuals. The threats to this species are not sufficiently understood, although in theory its inaccessibility should provide protection. However as the population occurs close to an inhabited area, its close proximity to human activities increases the risk of extinction and fire may also pose a threat.

## What is being done to protect it?

**Legally:** Though this species is included in the regional and national Red Lists, it is not protected by any law or convention.

*In situ*: The population occurs in the "Capo Gallo" Nature Reserve and SCI (Site of Community Importance) of "Capo Gallo" (ITA020006). This implies that collection of wild specimens is prohibited, grazing is regulated, and no quarrying is allowed.

*Ex situ*: Seeds are conserved in the germplasm bank of the Department of Botanical Science of Palermo University, and plants are cultivated in the botanical gardens of Palermo and Catania.

## What conservation actions are needed?

Fire management is key to the survival of this species and the preservation of the landscape of the nature reserve. Whilst fire is essential for some Mediterranean species which have co-evolved with fire and require it to survive, the increased frequency of human-made fire may pose a problem even to these plants and emphasises the importance of a carefully designed fire management scheme for the area.

## Scientific coordination

Professor Lorenzo Gianguzzi, Dipartimento Scienze Botaniche, Università degli Studi di Palermo, Italy.
Dr Antonino La Mantia, Dipartimento Scienze Botaniche, Università degli Studi di Palermo, Italy.

# *Petagnaea gussonei*

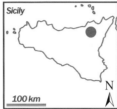

| | |
|---|---|
| Latin name: | *Petagnaea gussonei* (Sprengel) Rauschert |
| Synonym: | *Petagnia saniculaefolia* Guss. |
| Common name: | Falsa sanicola (Italian) |
| Family: | Umbelliferae (celery family) |
| Status: | ENDANGERED (EN) |

### Where is it found?

*Petagnaea gussonei* grows in the Nebrodi Mountains in the north-eastern part of Sicily at an altitude of 240-1,500 m. The species grows on the lower edge of the moisture-loving vegetation belts bordering shaded woodland mountain streams or rivulets which do not dry out in summer. The plant prefers sites where the current is calm and the substrate is soaked with water.

### How to recognise it

*Petagnaea gussonei* is a perennial herbaceous plant. The leaf has five equal lobes which are all the same size, with small teeth at the margins. The leaf stalk is attached to the centre of the leaf. Each inflorescence consists of four white flowers: a central stalkless flower, which can be either female or bisexual, surrounded by three stalked male flowers. These three are more or less united with the ovary of the central flower. The plant smells like common edible celery and is from the same family. *Petagnaea gussonei* usually flowers between May and June.

## Interesting facts

This species usually reproduces asexually through stolons; these horizontal branches grow out from the base of the plant and produce new plants from buds at their tips. This mechanism allows the plant to colonize the wet edges of streams. Seed production occurs occasionally, but the germination rate is very low.

## Why is it threatened?

This species has been categorized EN (Endangered) according to IUCN Red List Criteria B1ab(i,ii,iii,iv,v)+B2ab(i,ii,iii,iv,v). This is because there are only a few populations scattered over an area of approximately 400 km$^2$. In addition, it is believed that increased water use in the area will draw water away from the species habitat, resulting in a decline in area, habitat quality, number of populations and number of individuals.

The species was studied at the Department of Botanical Science of Palermo University for two years. The main threat identified was the reduction of water supply to the habitat by pumping and other human uses. Therefore, despite the fact that it grows in protected areas, if the water is removed this species will decline.

## What is being done to protect it?

**Legally:** Internationally, this species is listed in Appendix I of the Bern Convention and in Annexes II and IV of the EC Habitats Directive. In both these documents the species is listed under its synonym *Petagnia saniculifolia* Guss.

*In situ*: The known populations are mainly found in protected areas, including the Nebrodi Regional Park; the "Vallone Calagna sopra Tortorici" Nature Reserve; Sites of Community Importance (Torrente Fiumetto e Pizzo D'Ucina - ITA030002, Stretta di Longi - ITA030001); and Special Protection Zones (Serra del Re, Monte Soro e Biviere di Cesarò - ITA030038).

*Ex situ*: Seeds are conserved in a germplasm bank of the Department of Botanical Science of Palermo. The species is cultivated in the botanical gardens of Palermo, Catania and Messina on Sicily and at the English Gardens ("Giardini inglesi") of the Caserta Royal Palace ("Reggia di Caserta") on the Italian mainland just north of Naples.

## What conservation actions are needed?

More research is needed to identify all of this species' sites. Mountain water levels should be protected by prohibiting excessive use of mountain spring water.

The area of occupancy should be expanded by reintroducing the species in suitable sites within the area of occurrence, following *IUCN/SSC Guidelines For Re-Introductions*.

## Scientific coordination

Professor Lorenzo Gianguzzi, Dipartimento Scienze Botaniche, Università degli Studi di Palermo, Italy.
Dr Antonino La Mantia, Dipartimento Scienze Botaniche, Università degli Studi di Palermo, Italy.

# *Pleurotus nebrodensis*

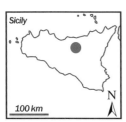

Sicily

100 km

N

| Latin name: | *Pleurotus nebrodensis* (Inzenga) Quélet |
|---|---|
| Common name: | **Funcia di basiliscu (Sicilian)** |
| Family: | **Pleurotaceae (Fungi Kingdom)** |
| Status: | **CRITICALLY ENDANGERED (CR)** |

### Where is it found?

*Pleurotus nebrodensis* is the only mushroom included in the "Top 50", and while technically not a plant, it was included to illustrate that some fungi as well as plants have been reduced to critically low levels and face imminent extinction. This mushroom only occurs in northern Sicily, growing in scattered localities in the Madonie mountains at an altitude of 1,200-2,000 m. It grows on limestone substrates, in pastures containing "Basiliscu" *Cachrys ferulacea*, a flowering plant belonging to the Umbelliferae or celery family.

### How to recognise it

The size of ripe fruiting bodies of this creamy to yellowish white mushroom varies. The diameter of its cap ranges from 5-20 cm, and is characterized by its deeply slanted whitish to pale yellow gills. Its stem is short and thick, either growing from the centre or to one side. When mature (approximately two or three days after the appearance on the ground of fruiting bodies), the surface of the cap breaks up. The species grows on rotten roots of *Cachrys ferulacea* in springtime from April to June.

## Interesting facts

*Pleurotus nebrodensis* was first described as *Agaricus nebrodensis* by Giuseppe Inzenga in 1863, and he called it "the most delicious mushroom of the Sicilian mycological flora". It has been a sought-after species since ancient times, and today remains a prized species. Given its rarity, it is sold in northern Sicily at a price fluctuating between 50-70 Euros per kilogramme. However, as the species is so rare, there is no formal market and the species is used in only a few restaurants where it is prepared using a number of traditional recipes.

## Why is it threatened?

This species has been categorized CR (Critically Endangered) according to IUCN Red List Criteria B1ab(iv,v)+2ab(iv,v), because the area where it is found is less than 100 km$^2$ and the population severely fragmented, and there is a decline in the number of localities and mature individuals. This is to due to the increasing number of mushroom gatherers, both professional and amateur, who are encouraged by the high price this mushroom commands. In addition to this increasing human pressure on the remaining natural populations, unripe fungi are usually collected. Due to the collection of unripe specimens, it is estimated that less than 250 individuals reach maturity each year.

## What is being done to protect it?

**Legally:** Currently, local regulations in the Madonie Park as well as a regional law do not exist. Draft rules have been prepared and submitted for the approval of the Government of Sicily. When approved, the collection of *Pleurotus nebrodensis* will be totally forbidden in zone A of the Park which is an integral reserve area. In other zones the collection of unripe mushrooms (i.e. those under 3 cm in diameter) will also be forbidden.

*In situ*: This species grows in Madonie Park which is a protected area. Experimental tests demonstrate that is possible to inoculate the roots of the host plant *Cachrys ferulacea* with the mushroom, thereby increasing mushroom production in the wild.

*Ex situ*: This species has started to be cultivated to reduce collection pressures on the species in the wild. The mushroom is grown in a tunnel made of metal arches varying in length between 20-30 m, and covered by a black net that provides 90% shade, and can be grown at various altitudes. Fortunately cultivated *Pleurotus* mushrooms retain the same characteristic aroma and flavour of the wild varieties, which is not the case with other species such as oyster mushrooms. *Ex situ* cultivation also provides additional income for local farmers, who can offer a cheaper product than that collected from the wild, which reduces the pressure on the wild population.

## What conservation actions are needed?

Legal action and enforcement is needed to stop over-collection and collection of unripe individuals of *Pleurotus nebrodensis* in the wild. At the same time the species needs to be cultivated *ex situ* to remove pressure on the wild populations. Reinforcement measures by inoculating the roots of its host plant and boosting wild production would also help, provided wild collection is carefully managed.

## Scientific coordination

Dr Giuseppe Venturella, University of Palermo, Italy.

# *Viola ucriana*

| | |
|---|---|
| **Latin name:** | *Viola ucriana* Erben & Raimondo |
| **Common name:** | Viola di Ucria (Italian) |
| **Family:** | Violaceae (violet family) |
| **Status:** | CRITICALLY ENDANGERED (CR) |

## Where is it found?

This perennial plant is only found on Mt. Pizzuta, near Palermo in north-western Sicily, growing at an altitude of 800-1,300 m. The only known population extends over two localities covering a total area of 0.2 km². The exact number of individuals is not yet known.

This little violet grows on eroded, sunny calcareous mountain slopes where rock outcrops or gravel substrates are sometimes covered by a garrigue vegetation characterized by the heath *Erica multiflora*, or more steppic vegetation dominated by the tall grass "Diss" (*Ampelodesmos mauritanicus*).

## How to recognise it

This perennial, evergreen herbaceous violet is partially hairy with greyish green leaves which are either heart- or sometimes inversely egg-shaped. The upper leaves are somewhat elongated while the lower leaves are gathered together at the base, forming a cushion. The flowers are yellow with a straight or slightly curved, yellowish-green spur. The species usually flowers between April and June.

## Interesting facts

This species only reproduces by seed. There are over 400 species of violets (*Viola* spp.) in the world, with more than 90 found in Europe. Taxonomic opinion on the species in Sicily has been divided, with some botanists "lumping" this species into the more common species *Viola nebrodensis*. However it is now generally recognised that *Viola nebrodensis* is divided into three separate species with localized distribution: *Viola ucriana* Erben & Raimondo (Pizzuta Mt.), *Viola tineorum* Erben & Raimondo (Rocca Busambra), and *Viola nebrodensis* Presl (Madonie Mts.).

## Why is it threatened?

This species has been categorized CR (Critically Endangered) according to IUCN Red List Criteria B1ab(ii,iii,v)+2ab(ii,iii,v). This is due to the extremely restricted area in which the single population is found, and the fact that the area, quality of habitat, and number of individuals are predicted to decline.

The area where *Viola ucriana* persists is often subject to fire due to human activities, and these periodic sumer fires pose a major threat to the conservation of this species. Forestry plantations of exotic conifers and other species are changing the ecological conditions of its habitat, and any human action could exterminate the one remaining population.

## What is being done to protect it?

**Legally:** This species is included in Red Lists at different levels (regional, national and international), either as part of the taxon *Viola nebrodensis* or as a variety of the latter. However, it is not protected by any particular law or convention.

*In situ:* The entire area where it is found is included in the "Monte Pizzuta, Costa del Carpineto, Moarda" SCI (Site of Community Importance). However, only the Mt. Pizzuta site is part of "Serre della Pizzuta" Nature Reserve.

*Ex situ:* Seeds are conserved in the germplasm bank of the Department of Botanical Science, Palermo University, and cultivated in the Botanical Garden of Palermo.

## What conservation actions are needed?

It is important to immediately stop forestry plantations in the species' habitat to limit the ecological transformation of the area. A protection scheme against fire is needed for the relevant sites. Ideally the entire population should be included in the Nature Reserve.

## Scientific coordination

Professor Lorenzo Gianguzzi, Dipartimento Scienze Botaniche, Università degli Studi di Palermo, Italy.
Dr Antonino La Mantia, Dipartimento Scienze Botaniche, Università degli Studi di Palermo, Italy.

# Zelkova sicula

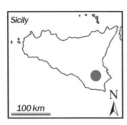

Sicily

100 km

N

| | |
|---|---|
| Latin name: | *Zelkova sicula* Di Pasquale, Garfì & Quézel |
| Common name: | Zelkova siciliana (Italian) |
| Family: | Ulmaceae (elm family) |
| Status: | CRITICALLY ENDANGERED (CR) |

### Where is it found?

Discovered in 1991, some 200-250 individuals of *Zelkova sicula* exist, growing in a very old cork oak (*Quercus suber*) forest. This single remote population extends 200 m along the banks of a stream on the northern slopes of the Iblei Mountains in south-eastern Sicily.

### How to recognise it

The tree species has a bushy habit and usually grows to 2-3 m tall. Its greyish bark is smooth on the young twigs, and tends to turn brownish-grey and peel off in plates on older trees. Leaves are oval shaped, bear teeth on the edges, and are densely covered by bristly hairs on both sides of the leaf. They are symmetric at their base (unlike many other members of the elm family), and are shed every autumn. The flowering period is in April.

Male and female flowers, as well as bisexual flowers, can be found on the same individual. No pollination has so far been observed, and most of the seeds produced are sterile. Consequently most of the reproductive success is due to root suckers, which are new stems that originate from the root system. Individuals produced in this manner are genetically identical to the original plant. Preliminary studies indicate that the remaining population probably goes back to just one individual (or at most a few). The vegetative propagation mode of *Zelkova* makes it difficult to determine the exact number of functionally independent individuals in the population.

This species has evolved in a Mediterranean climate with pronounced dry periods during summer. In extreme years, such conditions can lead to the withering of branches or of entire individuals, but a second regrowth can take place at the end of summer or beginning of autumn.

### Interesting facts

The genus *Zelkova* only has a few species, of which a few grow in Western and Eastern Asia, plus two in the Mediterranean basin (this species in Sicily and another, *Zelkova abelicea*, which is a threatened endemic from Crete). These two species represent relicts that have persisted over a long period of time and may have been more widespread in the past. Both are threatened by habitat change.

### Why is it threatened?

This species has been classified CR (Critically Endangered) according to IUCN Red List Criteria B1ab(i,ii,v)+2ab(i,ii,v); C2a(ii), D. This means that the area in which this species is found is very small, and both the area as well as the number of individuals continue to decline. The small number of individuals in a single declining population makes the probability of extinction of this species very high.

The ancient oak forest hosting this species is rather degraded due to overexploitation (logging, grazing, and removal of cork) for several centuries. In addition, over the last few years several major droughts have caused the death of several trees. *Zelkova sicula* requires relatively moist conditions. Therefore, if rainfall remains low, this species is almost certain to become extinct.

### What is being done to protect it?

**Legally:** Currently, there is no legal protection for this species.

*In situ*: The area where this species is found is fenced to prevent grazing. It is entirely situated within an SCI (Site of Community Importance) area, which is part of the European Natura 2000 network.

*Ex situ*: This species is cultivated at the Botanical Garden of the University of Catania, the Arboretum "Monna Giovannella" of the University of Florence, and at the Botanical Conservatory of Brest.

### What conservation actions are needed?

A restoration plan including both *ex situ* and *in situ* activities needs to be established. In order to determine the degree of affiliation between all the individuals in the population, a broader genetic study needs to be undertaken.

### Scientific coordination

Dr Giuseppe Garfì, CNR – Instituto per i Sistemi Agricoli e Forestali del Mediterraneo (ISAFoM), Sezione di Ecologia ed Idrologia Forestale, Rende, Italy.

# *Centaurea gymnocarpa*

Tuscan Archipelago

N
30 km

Capraia

N
3 km

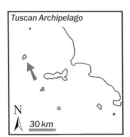

| | |
|---|---|
| **Latin name:** | *Centaurea gymnocarpa* Moris & De Not. |
| **Common name:** | **Fiordaliso di Capraia (Italian)** |
| **Family:** | **Compositae (daisy family)** |
| **Status:** | **ENDANGERED (EN)** |

### Where is it found?

*Centaurea gymnocarpa* is endemic to the Italian island of Capraia, a small island which is part of the Tuscan Archipelago. It is a species that colonizes cracks and fissures in rock faces, growing on acid rocks. It is found in association with *Linaria capraia* and other endemic species such as *Silene badaroi* and *Galium caprarium*. The localities of *Centaurea gymnocarpa* have been described as Habitat 8220 in the EC Habitats Directive.

## How to recognise it

*Centaurea gymnocarpa* is a herbaceous plant with a woody base and a long herbaceous stem, which can reach 80-100 cm in height. It is covered with dense short hairs that gives it a greyish-white colour. The leaves, more or less fleshy, are green above and greyish-white below, and many old leaves persist at the base. Its tiny pink flowers are gathered together in a compact flower head. Flowering occurs in May.

## Interesting facts

*Centaurea gymnocarpa* belongs to the group known as the *"cineraria"* group. This group probably was once a single species when the land masses were united, but as islands were formed, new species evolved on each island. This means that today there are a number of closely related species of *Centaurea* in the Mediterranean growing on rocky seaward cliffs, all probably related to a common ancestor.

## Why is it threatened?

This species has been categorized EN (Endangered) according to IUCN Red List Criterion D, as the population size is estimated to be less than 250 individuals. The decline in this small population is not yet significant enough to classify this species as Critically Endangered, as only one subpopulation is under threat. This subpopulation, situated between Paese and Porto, is severely threatened by competition from two invasive plants: *Carpobrotus acinaciformis* and *Senecio angulatus*. These two species in the last years have been expanding in Capraia, but do not yet grow in the other areas where *Centaurea gymnocarpa* occurs, which are for the most part far away from areas of human habitation.

Eight subpopulations (censused in 1x1 km grids) have been identified, covering less than 10 km$^2$. The one threatened subpopulation is composed of less than 20 individuals, while the seven others include more then 20 individuals.

## What is being done to protect it?

**Legally:** This species is protected by the law 56/2000, which is a law guiding biodiversity conservation in the Tuscan region, and is quite similar to the EC Habitats Directive. Under this law, it is forbidden to collect any species in this genus.

*In situ*: Four of the eight known subpopulations occur in the Tuscan Archipelago National Park. The Park includes a protected terrestrial area of just under 18,000 ha, and a marine protected area of approximately 60,000 ha (making it the largest European marine park). The objective of the Tuscan Archipelago National Park is to protect this fragile natural environment, which is culturally and scientifically very rich.

*Ex situ*: The species is cultivated in the Botanical Garden of Florence.

## What conservation actions are needed?

Monitoring of all subpopulations is needed, and a programme to remove invasive alien plants which threaten one of the subpopulations needs to be undertaken. Efforts to ensure that these alien species do not start growing in the other areas where this species is found are also very important. Once the alien species have been eradicated, a re-introduction programme will be planned, using specimens propagated from the threatened population.

## Scientific coordination

Dr Bruno Foggi, Dipartimento Biologia Vegetale, Università di Firenze, Firenze, Italy.

# Glossary

**Alternate:** leaves placed alternately along the stem, not opposite.

**Annual:** a plant that completes its entire life cycle from seed to flower to seed again within one year.

**Bern Convention:** an international convention which aims to promote European co-operation in conserving wild flora and fauna and their habitats. States undertake legislative and administrative measures to protect the wild flora species specified in Appendix I, which includes prohibiting the deliberate picking, collecting, cutting or uprooting of such plants.

**Biennial:** plants which live for two years. Usually the first year's growth produces a leaf-rosette, the second the flowers.

**Boreal:** related to northern regions.

**Bract:** specialized leaf or leaf-like part, usually situated at the base of a flower or inflorescence.

**Calcareous soils:** those formed on calcium carbonate rich rocks such as limestone or chalk. Lime-rich soils have a different and usually richer association of plants than acid soils.

**Calyx:** refers to the sepals (sterile parts of the flower inserted below the petals) as a whole, usually when they are joined.

**Capsule:** a dry fruit that when mature splits apart to release the seeds within.

**Compound:** for leaves, this means that the leaf is separated into leaflets. An inflorescence can also be compound, meaning that it is branching.

**Corolla:** refers to the petals as a whole, usually when they are joined.

**Elliptical:** shape of leaf or leaflet which is widest at the middle.

**Endemic:** a species native to a particular region, such as only native to an island, a mountain, or a country.

**EC Habitats Directive:** a Directive adopted by the Member States of the European Union to help maintain biodiversity by defining a common framework for the conservation of wild plants and animals and habitats of Community interest. Annex II (animal and plant species of Community interest) to the Directive lists the habitats and species whose conservation requires the designation of special areas of conservation. Some of them are defined as "priority" habitats or species (in danger of disappearing). Annex IV lists animal and plant species in need of particularly strict protection.

**Garrigue:** an open, shrubby, evergreen Mediterranean vegetation, usually occurring on calcareous soils, resulting from forest regression due to fire and intensive grazing.

**Genotype:** the specific genetic makeup of an individual.

**Gill:** the spore-bearing, radiating structures found underneath certain mushroom caps.

**Glabrous:** having no hair or similar growth.

**Glandular hairs:** short or long hairs with a gland at the tip, seen as a swelling or blob, and often giving the plant a sticky feel.

**Hispid:** having stiff coarse hairs or bristles.

**Inflorescence:** a group or cluster of flowers.

**Invasive species:** organisms which successfully establish themselves in, and then overcome, otherwise intact, pre-existing native ecosystems. The consequences of such invasions, including alteration of habitat and disruption of natural ecosystem processes, are often catastrophic for native species as well as for human livelihoods. An alien (non-native or exotic) species is one occurring outside of its natural range that has been directly or indirectly introduced by people.

**Lanceolate:** leaf or leaflet which is spear-shaped; a narrow leaf broader at the base and tapering to a point.

**Linear:** shape of leaf or leaflet which is long and narrow, almost parallel-sided.

**LIFE Project:** LIFE (The Financial Instrument for the Environment) co-finances environmental initiatives in the European Union and certain third countries.

**Leaf axil:** the point at which the leaf stem is attached to a stem or branch.

**Leaflet:** a division or part of a compound leaf.

**Mycorrhiza:** a mutually beneficial (symbiotic) association between a plant root and a fungus that enhances the ability of the root to absorb water and nutrients.

**Natura 2000:** a network established by the EC Habitats Directive. This network comprises "special areas of conservation" designated by Member States in accordance with the provisions of the Directive, and "special protection areas" classified pursuant to Directive 79/409/EEC on the conservation of wild birds.

**Opposite:** of leaves arising opposite each other on the stem, thus appearing in pairs.

**Perennial:** plants that persist for many growing seasons. Often the top portion of the plant dies back during winter or the dry season and regrows from the same root system, although many perennial plants keep their leaves year round.

**Pod:** a fruit, usually long, cylindrical and never fleshy, as in peas.

**Population:** in the text of this booklet, population has been used in the biological sense as a community of individuals sharing a common gene pool. Subpopulations are groups of these individuals that are isolated geographically. However note that the term population is used differently in the IUCN Red List Categories and Criteria, which defines population as the total number of mature individuals of the taxon.

**Relict:** an organism that at an earlier time was abundant in a large area but due to some major change (such as climatic or land use) is now occurring at only one or a few small areas.

**Rhizome or rootstock:** a horizontal, underground stem of a plant, generally modified (particularly for storing food materials), that often sends out roots and shoots from its nodes.

**Rosette:** A flattened, rose-like group of leaves at the base of a stem.

**Spanish Royal Decree 439/1990:** this Decree, of 30 March 1990, regulates the National Catalogue of Endangered Species. The species included in Annex I (species in danger of extinction) are the object of a recovery plan.

**Speciation:** the evolutionary formation of new biological species, usually by the division of a single species into two or more genetically distinct ones.

**Spur:** a hollow, tubular extension to a petal in some flowers, often containing nectar.

**Stalk:** the slender stem that supports a leaf or a flower.

**Stamen:** the male reproductive organ of a flower that produces pollen.

**Stolon:** similar to a rhizome, but exists above ground, sprouting from an existing stem.

**Succulent:** a plant adapted to arid conditions and characterized by fleshy water-storing tissues that act as water reservoirs.

**Taxonomy:** the science of classifying living organisms.

# Selected references

Due to space limitations, only selected references have been included in this booklet. However, a complete list of references for all the Top 50 information is included on the Top 50 website, see www.iucn.org/themes/ssc/plants/top50/

## General publications on species conservation

Davis, S.D., Heywood, V.H. & Hamilton, A.C. (Eds). 1994–1997. *Centres of Plant Diversity. A Guide and Strategy for Their Conservation.* 3 vols. World Wide Fund for Nature, Gland, Switzerland and IUCN, Gland, Switzerland and Cambridge, U.K.

Given, D.R. 1994. *Principles and Practice of Plant Conservation.* Timber Press, Portland, Oregon, USA. 292 pp.

Guerrant, E.O., Havens, K., & Maunder, M. (Eds). 2004. *Ex Situ Plant Conservation: Supporting Species Survival in the Wild.* Island Press, USA. 504 pp.

IUCN. 2001. *IUCN Red List Categories and Criteria: Version 3.1.* IUCN Species Survival Commission, IUCN, Gland, Switzerland and Cambridge, U.K. ii + 30 pp.

IUCN. 2004. *IUCN Red List of Threatened Species.* URL: http://www.iucnredlist.org. [The IUCN Red List is updated annually].

Mittermeier, R.A., Robles Gil, P., Hoffman, M., Pilgrim, J., Brooks, T., Goettsch Mittermeier, C., Lamoreux, J. & da Fonseca, G.A.B. 2004. *Hotspots Revisited: Earth's Biologically Richest and Most Threatened Terrestrial Ecoregions.* Conservation International, Washington, D.C., USA. 390 pp.

## Specific publications on Mediterranean flora and islands

Alomar, G., Mus, M. & J.A. Rosselló. 1997. *Flora endèmica de les Balears.* Consell Insular de Mallorca, Palma de Mallorca, Spain.

Alonso, L.A., Carretero, J.L. & Garcia-Carrascosa, A.M. (Eds). 1987. *Islas Columbretes: contribución al estudio de su medio natural.* Monografias, n° 5, Generalitat Valenciana, Valencia, Spain.

Bañares, Á., Blanca, G., Güemes, J., Moreno J.C. & Ortiz, S. (Eds). 2003. *Atlas y Libro Rojo de la flora vascular amenazada de España: taxones prioritarios.* Dirección General de Conservación de la Naturaleza. Madrid, Spain.

Blamey, M. & Grey-Wilson, C. 2004. *Mediterranean Wild Flowers.* Domino Books / A & C Black, U.K. 560 pp.

Blondel, J. & Aronson, J. 1999. *Biology and Wildlife of the Mediterranean Region.* Oxford University Press, New York, USA. 352 pp.

Castroviejo, S. (Coord.). 1986–. *Flora Iberica: plantas vasculares de la Península Ibérica e Islas Baleares.* Real Jardín Botánico, Madrid, Spain. URL: http://www.rjb.csic.es/floraiberica/

Conti, F., Manzi, A. & Pedrotti, F. 1992. *Libro Rosso delle piante d'Italia.* Società Botanica Italiana and World Wildlife Fund Italian Association. Camerino (MC). Roma, Italy. 637 pp.

Danton, P. & Baffray, M. 1995. *Inventaire des plantes protégées en France.* Yves Rocher, AFCEV, Nathan, France. 234 pp.

Delanoë, O., Montmollin, B. de, Olivier, L. & the IUCN/SSC Mediterranean Islands Plant Specialist Group. 1996. *Conservation of Mediterranean Island Plants. 1. Strategy for Action.* IUCN, Gland, Switzerland and Cambridge, U.K. 106 pp.

Dominguez Lozano, F. (Ed.). 2000. Lista Roja de la flora vascular española – Red List of Spanish vascular flora. *Conservación Vegetal* (extra) 6. 39 pp.

Gamisans, J. & Jeanmonod, D. 1993. *Catalogue des plantes vasculaires de la Corse*. 2nd edition. In: D. Jeamonod & H.M. Burdet (Eds), *Compléments au Prodrome de la flore de Corse*. Conservatoire et Jardin botaniques de Genève, Switzerland. 258 pp.

Gamisans, J. & Marzocchi J.-F. 1996. *La flore endémique de la Corse*. EDISUD, Aix-en-Provence, France. 208 pp.

Goméz-Campo, C. (Ed.). 1985. *Plant Conservation in the Mediterranean area*. Dr. W. Junk, Dordrecht, Germany.

Goméz-Campo, C. (Coord.). 2001. *Conservación de especies vegetales amenazadas en la región Mediterránea occidental. Una perspectiva el fin de siglo*. Fundación Ramón Areces, Madrid, Spain.

Greuter, W., Burdet, H.M. & Long, G. (Eds.) 1984; 1986; 1989. *Med-Checklist. A Critical Inventory of Vascular Plants of the Circum-Mediterranean Countries*. Vol. 1, 3 & 4. Geneva, Switzerland & Berlin, Germany.

Haslam, S.M., Sell, P.D. & Wolseley, P.A.W. 1977. *A Flora of the Maltese Islands*. University Press, Malta. lxxi + 560 pp.

Laguna, E. & Jimenez, J.L. 1995. Conservación de la flora de las islas Columbretes. *Ecologia Mediterranea* 21(1 & 2): 325–336.

Lanfranco, E. 1977. *A Field Guide to the Wild Flowers of Malta*. Progress Press, Malta. viii + 83 pp. + 65 plates.

Lanfranco, E. 1989. The Flora. In: P.J. Schembri & J. Sultana (Eds), *Red Data Book for the Maltese Islands*. Department of Information, Malta. Pp. 5–70.

Meikle, R.D. 1977, 1985. *Flora of Cyprus*. The Bentham - Moxon Trust Royal Botanic Gardens, Kew, U.K.

Mota, J.F., Sola, A.J., Aguilera, A., Cerrillo, M.I. & Dana, E. 2002. The Mediterranean Island of Alborán: a review of its flora and vegetation. *Fitosociologia* 39(1): 15–21.

Olivier, L., Galland, J.-P., Maurin, H. & Roux, J.P. 1995. *Livre Rouge de la flore menacée de France. Tome 1: espèces prioritaires*. Collections patrimoines naturels, Muséum National d'Histoire Naturelle, Cons. Botanique Nat. de Porquerolles & Ministère de l'Environnent, Paris, France. 486 pp.

Olson, D. & Dinnerstein, E. 1998. The Global 200: A representative approach to conserving the Earth's most biologically valuable ecoregions. *Conservation Biology* 12:502-515.

Pantelas, V., Papachristophorou, T. & Christodoulou, P. 1993. *Cyprus Flora in Colour: The Endemics*. Nicosia, Cyprus.

Phitos, D., Strid, A., Snogerup, S. & Greuter, W. (Eds). 1995. *The Red Data Book of Rare and Threatened Plants of Greece*. WWF-Greece, Athens, Greece. 527 pp.

Pignatti, S. 1982. *Flora d'Italia*. 3 vols. Edagricole, Bologna, Italy. 780 pp.

Raimondo, F.M., Gianguzzi, L. & Ilardi, V. 1994. Inventario delle specie "a rischio" nella flora vascolare nativa della Sicilia. *Quad. Bot. Ambientale Appl.* (1992) 3: 65-132.

Tan, K. & Iatroú, G. 2001. *Endemic Plants of Greece. The Peloponnese*. Gad Publishers Ltd., Copenhagen, Denmark. 480 pp.

Tutin, T.G., Heywood, V.H., Burges, N.A., Chater, A.O., Edmonson, J.R., Moore, D.M., Valentine, D.H., Walters, S.M. & Webb, D.A. (Eds). 1993. *Flora Europaea*. 5 vols. Cambridge University Press, U.K.

Valsecchi, F. 1980. Le piante endemiche della Sardegna: 80–83. *Boll. Soc. Sarda Sci. Nat.* 19: 323–342.

Zervakis, G. & Venturella G. 2002. Mushroom breeding and cultivation enhances *ex situ* conservation of Mediterranean *Pleurotus* taxa. In: J.M.M. Engels, V.R. Rao, A.H.D. Brown & M.T. Jackson. (Eds), *Managing Plant Genetic Diversity*. CABI Publishing, U.K. Pp. 351–358.